Understanding Revelations by Astronomy

The Ancient, World-Wide, Hidden Knowledge,
Especially in Genesis & Revelations;
on Precessional Cosmology,
the Galactic Alignment, & Sacred Geometry.

Nemo Tanis Lassiter

"If seven maids with seven mops swept for half a year…"
From the poem, *The Walrus and the Carpenter*,
in *Through the Looking Glass*,
by Lewis Carroll.

1st paperback edition… 108 pages
copyright 2017;
ISBN: 9781549847776
Independently published

This text in its entirety is the sole possession of Nemo Tanis Lassiter. All rights reserved. No part of this text may be reproduced or transmitted in any form or by any means; verbal, electronic or mechanical; including photocopying, recording, or by any information storage and retrieval system, without permission in writing from the author; excepting what is entitled by the 'fair use' act.

Acknowledgements:

I wish to thank my beloved parents, Matt and Cathy, who smile on me from the divine. And I am most thankful for my daughter and only child, Cathy, whom I cherish above all else. I must give special thanks to Katherine Dunham and Sir Laurence Gardner, as kindred spirits and loved ones I have met in person.

I also wish to thank my ancestors, and the spiritual forces that have guided me and shown me wonders. Thank you- God the Most High, Holy Sophia, and Christ who holds the door to Heaven.

I am a seeker of knowledge and it would take too long to list the authors who have influenced me with their books, songs and movies. I will only mention *Hamlet's Mill,* written by two masters of Celestial Mythography…

Hertha Von Dechend, the greater of the two according to her coauthor, Giorgio De Santillana, who did more public speaking and lectures due to Hertha's less precise command of the English language.

I use the term Precessional Cosmology over the much more common term, Archeo-Astronomy. This academic field is split in twain. The book I refer too so much in my work, *Hamlet's Mill,* is a masterwork of Celestial Mythography. Authors, such as myself and others, value it almost as a sacred text. However, many within Archeo-Astronomy refuse to admit its validity and believe the stellar encoding of the mythos by the ancients to be absolutely false. While the edifices and monumental remains of our forebearers certainly exist, they are not the whole of this academic principal. As such, it is inaccurate to name the entire field, from only one of its three disciplines.

1) Archeo-Astronomy. 2) Celestial Mythography. 3) the Stellar Mechanism

It would also be a very long list for me to name the friends and especially the kindred I have made in my life of travels, but they know who they are. And it is my hope that they remember me fondly sometimes when they gaze up at the starry sphere.

Please send any communications to me via email:

precessionalcosmology@yahoo.com

I remain,

Nemo Tanis Lassiter

O Quam Misericours Est Deus, Justus Et Pius ?

Table of Contents:

Forward, An Invitation for Graham Hancock .... p. 1

Prologue: A Bigger Picture .... p. 3

Part I: The Visions of Saint John…

- Ch 1: Revelations XII .... p. 6
- Ch 2: Precessional Astronomy Basics .... p. 7
- Ch 3: The Celestial & the Ecliptic .... p. 7
- Ch 4: The Celestial Axis Swivels around the Ecliptic .... p. 9
- Ch 5: The EYE of God, Heaven, Horus & Sauron .... p. 11
- Ch 6: The EYE Moves .... p. 11
- Ch 7: The Galactic Alignment .... p. 13
- Ch 8: The Year of the Messiah's Birth .... p. 14
- Ch 9: The Crown of 12 Stars .... p. 16
- Ch 10: The Galactic Center .... p. 18
- Ch 11: Dividing the 'Great Year' .... p. 19
- Ch 12: The Three Dragons, 1) The Hydra .... p. 24
- Ch 13: The Three Dragons, 2) Draco .... p. 25
- Ch 14: The Constellations of the Beast .... p. 26
- Ch 15: The Beast & the Woman with a Crown .... p. 30
- Ch 16: The Three Dragons, 3) Ophiuchus .... p. 31
- Ch 17: The Serpent's Tail & the Eagle's Wing .... p. 34
- Ch 18: The Male-Child Messiah .... p. 34
- Ch 19: The Shield of Kings .... p. 37
- Ch 20: Who is the Serpent Bearer? .... p. 39
- Ch 21: The Rod of Iron .... p. 40
- Ch 22: The Great Mystery of 108 .... p. 47

Part II: Sacred Geometry and Precession

Ch 23: Seventh Heaven .... p. 51

Ch 24: The 14-Fold Division .... p. 53

Ch 25: The Circled-Square & Cubed-Square .... p. 54

Ch 26: The Beast and the Dragon with the 5-4-5 Triangle .... p. 55

Ch 27: *The Giza Pyramids of Egypt .... p. 58

Ch 28: The Pyramids at Chichen Itza .... p. 58

Ch 29: *Copernicus, Chichen Itza, Stonehenge, & Washington D.C. .... p. 60

Ch 30: The Galactic Axis and the Ecliptic Axis .... p. 65

Ch 31: Sighted Down the Earth's Celestial Pole .... p. 66

Ch 32: Once again, the Sacred Geometry .... p. 69

Ch 33: The 5-4-5 Split of the Calendar .... p. 70

Ch 34: The 4-3-4-3 'X' Shape and the Calendar .... p. 72

Part III – The Greater Alignments of Precession

Ch 35: A Review So Far .... p. 74

Ch 36: The Full Celestial Cycle .... p. 79

Ch 37: The Full Ecliptic Cycle .... p. 80

Ch 38: January 1st & July 3rd .... p. 81

Ch 39: Jupiter & it's Invariable Plane .... p. 82

Ch 40: The Shape of the Earth's Orbit .... p. 83

Ch 41: And Now, All of it Together .... p. 85

(Part II has a review before starting Part III. Then, after the new information is presented in Part III; there is a short summary of the entire section on the Sacred Geometry and the Greater Alignments.)

Part IV – The Art of the Fugue

- Ch 42: The Dictionary Definition .... p. 93
- Ch 43: *Hamlet's Mill* .... p. 94
- Ch 44: The Fugue is Older than Dirt .... p. 94
- Ch 45: The Visual Story .... p. 95
- Ch 46: The Jam of Preserves .... p. 96
- Ch 47: The Fugue Gets Lost & Found .... p. 96
- Ch 48: The Renaissance of the FUGUE .... p. 98
- Ch 49: Recognizing the Puzzle Pieces .... p. 98
- Ch 50: The Qualities of a Pirate .... p. 99
- Ch 51: The Storyteller's Tale .... p. 104
- Ch 52: The Elephant in the Room .... p. 106

Appendix: The Science of Astronomy .... p. 107

List of Figures:

* These figures are all in the Public Commons.

\+ These figures have been altered 3 ways from their source. All rights are reserved.

All other figures are the author's creation. All right are reserved.

PART I: The Visions of Saint John

Fig. 1: *Medieval Star Chart, North Ecliptic Pole #1 .... p. 8

Fig. 2: *Medieval Star Chart, North Celestial Pole #1 .... p. 8

Fig. 3: Ecliptic Armillary #1 .... p. 8

Fig. 4: Ecliptic Armillary #2 .... p. 9

Fig. 5: Circles of Precession #1 .... p. 10

Fig. 6: Ecliptic Armillary #3 .... p. 10

Fig. 7: Hemisphere Ecliptic Armillary #1 .... p. 10

Fig. 8: The EYE – Hemisphere Ecliptic Armillary #2 .... p. 11

Fig. 9: Hemi. Ecliptic Armillary #3 & Hemi. Ecl. Arm. W- Galactic Equator #1 .... p. 12

Fig. 10: Hemi. Ecliptic Armillary w- Galactic Equator #2 .... p. 12

Fig. 11: Ecliptic Equator w- Milky Way #1 .... p. 13

Fig. 12: +Cylinder Projection, Ecliptic Armillary w- Galactic Equator #1 .... p. 13

Fig. 13: +Cylinder Proj., Ecl. Armillary w- Gal. Equator #2 .... p. 14

Fig. 14: +Leonids Meteor Shower #1 .... p. 15

Fig. 15: +Leonids Meteor Shower #2 .... p. 16

Fig. 16: Ecliptic Equator w- Milky Way #2 .... p. 18

Fig. 17: Galactic Center w- Galactic Anti-Center #1 .... p. 18

Fig. 18: +Zodiac Constellations and Signs #1 .... p. 19

Fig. 19: World Ages of Metal #1 .... p. 20

Fig. 20: Constellations, Signs, w- Ages of Metal #1 .... p. 21

Fig. 21: Constellations, Signs, w- Ages of Metal #2 .... p. 21

Fig. 22: Constellations, Signs, w- Ages of Metal #3 .... p. 22

Fig. 23: Constellations, Signs, w- Ages of Metal #1 .... p. 23

Fig. 24: *Caprarola Fresco, the Hydra #1 .... p. 24

Fig. 25: +The Hydra #2 .... p. 25

Fig. 26: *Hevellius' Star Chart, Draco #1 .... p. 26

Fig. 27: *Apian's Star Chart, Virgo #1... p. 28

Fig. 28: *Apian's Star Chart, Virgo #2... p. 29

Fig. 29: *Honter's Star Chart, Virgo #3 .... p. 29

Fig. 30: *Backer's Star Chart, Bernice's Hair #1 .... p. 30

Fig. 31: *Backer's Star Chart, Bernice's Hair #2 .... p. 30

Fig. 32: *Argelander's Star Chart, Bernice's Hair #3 .... p. 31

Fig. 33: *Apian's Star Chart, the Serpent Bearer #1 .... p. 32

Fig. 34: *Caprarola Fresco, the Serpent Bearer #2 .... p. 33

Fig. 35: *Backer's Star Chart, the Serpent Bearer #3 .... p. 33

Fig. 36: *Apian's Star Chart, the Serpent's Tail and the Eagle's Wing #1 .... p. 34

Fig. 37: *Backer's Star Chart, the Man-Child #1 .... p. 35

Fig. 38: *Caprarola Fresco, the Man-Child #2... p. 36

Fig. 39: *Hevellius' Star Chart, the Shield #1 .... p. 38

Fig. 40: *Argelander's Star Chart, the Shield #2 .... p. 38

Fig. 41: *Aratus' Star Chart, Gemini and Ophiuchus #1 .... p. 39

Fig. 42: *Aratus' Star Chart, Gemini and Ophiuchus #2 .... p. 40

Fig. 43: +Northern Hemisphere #1 .... p. 41

Fig. 44: +Northern Hemisphere #2 .... p. 41

Fig. 45: Milky Way Gal. Center, the Jaw Bone #1 .... p. 42

Fig. 46: Hemisphere Ecliptic Armillary #1 .... p. 42

Fig. 47: +Cylinder Proj., Ecl. Armillary w- Gal. Equator #1 .... p. 43

Fig. 48: +Cylinder Proj., Ecl. Armillary w- Gal. Equator #2 .... p. 43

Fig. 49: +Cylinder Proj., the Rod of Iron #1 .... p. 44

Fig. 50: *Apian's Star Chart, the Rod of Iron #2 .... p. 44

Fig. 51: *Hevellius' Star Chart, the Rod of Iron #3.. p. 45

Fig. 52: +The Rod of Iron #4 .... p. 46

Fig. 53: Circles of 360 Degrees & 108 Degrees #1 .... p. 50

vii

UNDERSTANDING REVELATIONS BY ASTRONOMY

PART II, Sacred Geometry & Precession

Fig. 54: *Hamlet's Mill Star Chart, North Celestial & Ecliptic Poles #1 ..... p. 51

Fig. 55: *Medieval Star Chart, North Celestial Pole #1 ..... p. 51

Fig. 56: 7-Fold Division, Concentric Circles & Lightening Spokes #1 ..... p. 52

Fig. 57: 1/7th of a Circle Measurement #1 ..... p. 52

Fig. 58: 14-Fold Division of a Circle, 7 & 7 Division of a Circle #1 ..... p. 53

Fig. 59: 4-3-4-3 'X' Division of a Circle, 5-4-5 Trine Division of a Circle #1 ..... p. 53

Fig. 60: Circled Square & the Cubed Sphere #1 ..... p. 54

Fig. 61: Masonic Checkerboard Cross #1 ..... p. 55

Fig. 62: *Medieval Star Chart, North Celestial Pole & North Ecliptic Pole #1 ..... p. 55

Fig. 63: +The Dragon and the Beast #1 ..... p. 56

Fig. 64: The Heads, Horns, & Crowns #1 ..... p. 57

Fig. 65: +Giza Pyramid & the 1/7th Angle #1 ..... p. 58

Fig. 66: +Giza Pyramid & the Airshafts #1 ..... p. 58

Fig. 67: Chichen Itza, Kukulcan's Pyramid #1 ..... p. 59

Fig. 68: Chichen Itza, Kukulcan's Pyramid #2 ..... p. 59

Fig. 69: Chichen Itza, High Priest's Pyramid #1 ..... p. 59

Fig. 70: *Copernicus' Manuscript # 1 ..... p. 60

Fig. 71: *Copernicus' Sun-Centered System #1 ..... p. 61

Fig. 73: +Chichen Itza, Sun Rise & Sun Set #1 ..... p. 62

Fig. 74: +Chichen Itza, Archeological Site #1 ..... p. 62

Fig. 75: +Chichen Itza, Archeological Site #2 ..... p. 63

Fig. 76: +Chichen Itza, Archeological Site #3 ..... p. 63

Fig. 77: +Chichen Itza, Archeological Site #4 ..... p. 63

Fig. 78: Pentagram & the Golden Ratio #1 ..... p. 63

Fig. 79: +Chichen Itza, The Plaza #1 ..... p. 64

Fig. 80: +Stonehenge & +Washington DC #1 ..... p. 64

Fig. 81: Circles of Precession #2 & 60 Degrees between Axis ..... p. 65

Fig. 82: The Shared Plane between the Galactic and Ecliptic Axis' #1 ..... p. 65

viii

UNDERSTANDING REVELATIONS BY ASTRONOMY

Fig. 88: The Shared Plane between the Galactic and Ecliptic Axis' #2 .... p. 66

Fig. 89: The Celestial, Ecliptic, & Galactic Axis' #1 .... p. 66

Fig. 90: The Galactic & Ecliptic Axis Sighted down the Celestial Axis #1 .... p. 67

Fig. 91: The Galactic & Ecliptic Axis Sighted down the Celestial Axis #2 .... p. 67

Fig. 92: The 4-3-4-3 'X' Sighted Down the Celestial Axis #1 .... p. 68

Fig. 94: The 4-3-4-3 'X' of the Galactic & Ecliptic Axis #1 .... p. 68

Fig. 95: The Circled-Square, Perimeter & Area #1 .... p. 69

Fig. 96: The Cubed-Sphere, Surface & Volume #1 .... p. 70

Fig. 97: The 5-4-5 split of the Calendar Year #1 .... p. 71

Fig. 98: The 4-3-4-3 'X' & the 5-4-5 split of the Galactic & Ecliptic Axis #1 .... p. 72

Fig. 99: The 4-3-4-3 'X' of the New Year and the Equinoxes #1 .... p. 72

PART III – The Greater Alignments of Precession

Fig. 100: +Atlas Holds the Heaven's #1 .... p. 74

Fig. 102: The Circled-Square, Perimeter & Area #1 .... p. 75

Fig. 103: The Cubed-Sphere, Surface & Volume #1 .... p. 75

Fig. 104: Masonic Checkerboard Cross #1 .... p. 76

Fig. 105: The 4-3-4-3 'X' & the 5-4-5 split of the Gal. & Ecl. Axis' #1 .... p. 77

Fig. 106: +The Dragon and the Beast #1 .... p. 77

Fig. 107: The Beast & the Dragon, Heads, Horns, & Crowns #1 .... p. 78

Fig. 108: The Full Celestial Tilt Cycle #1 .... p. 79

Fig. 109: The Full Celestial Tilt Cycle #2 .... p. 79

Fig. 110: The Full Ecliptic Tilt Cycle #1 .... p. 80

Fig. 111: The Full Ecliptic Tilt Cycle #2 .... p. 80

Fig. 112: +The Gal. and Cel. Equators' & 4-3-4-3 'X' shape #1 .... p. 81

Fig. 113: Jupiter, it's Planetoids & Invariable Plane #1 .... p. 82

Fig. 114: Jupiter's Invariable Plane Crossing the Ecliptic Plane #1 .... p. 83

Fig. 115: The Shape of the Earth's Orbit #1 .... p. 84

Fig. 116: The Shape of the Earth's Orbit #2 .... p. 84

Fig. 117: The Shape of the Earth's Orbit #3 .... p. 85
Fig. 118: Circled Square & the Cubed Sphere #1 .... p. 85
Fig. 119: Masonic Checkerboard Cross #1 .... p. 86
Fig. 120: 4-3-4-3 'X' & the 5-4-5 trine Division of a Circle #1 .... p. 86
Fig. 121: +The Dragon and the Beast, Heads, horns, & Crowns #1 .... p. 87
Fig. 122: The Celestial, Ecliptic, & Galactic Axis' #1 .... p. 87
Fig. 123: The 4-3-4-3 'X' & the 5-4-5 Split Hemisphere Charts #2 .... p. 88
Fig. 124: The 5-4-5 split of the Calendar, & Hemi. +The Gal. & Cel. Equators #2 .... p. 88
Fig. 125: The Full Celestial Tilt Cycle & The Full Ecliptic Tilt Cycle #1 .... p. 89
Fig. 126: +Hemi. The Gal. & Cel. Equators & Invariable Crossing the Ecl. #1 .... p. 89
Fig. 127: The Shape of the Earth's Orbit #4 .... p. 89
Fig. 128: The Shape of the Earth's Orbit #1 .... p. 90
Fig. 129: +Hemi. The Gal. & Cel. Equators & the 4-3-4-3 'X' Shape #1 .... p. 90
Fig. 130: The Cycles Around Jan. 1st & Jul. 3rd #1 .... p. 91

Forward, an Invitation to Graham Hancock:

At first, this space was reserved for the Archeo-Astronomer John Major Jenkins. His phenomenal book *Maya Cosmogenesis 2012*, was my introduction to the field of Archeo-Astronomy / Precessional Cosmology. And I was there at Chichen Itza for the March Equinox and the May 20th solar eclipse, in 2000. As well as, the December Solstice of 2012 and 2013. I will be forever grateful to him as a scholar and pioneer. It was John Major Jenkins who wrote the introduction to the edition of *Hamlet's Mill* I purchased. I read it after encountering his high regard of it. Unfortunately, John Major was taken early by cancer and is reported to have crossed over into the divine on 2 July 2017. This book is dedicated to him, for being the first breadcrumb on the trail I followed.

So now... These first 2 pages of my text are reserved for an introduction by Graham Hancock, whom I also have such a high regard for... I had originally planned on having Graham Hancock write the afterwards to this text, but now I have moved my appendix to the place of the afterward. And so, I offer an invitation to Graham Hancock alone. Please read this text when you find out about it, and if it rocks your world as much your work has rocked mine, then I ask you to write my forward.

UNDERSTANDING REVELATIONS BY ASTRONOMY

Prologue: A Bigger Picture

Revelations is understood to give clues about what's called the SECOND COMING OF CHRIST. This event is said to be the "End of Times". And, while most people believe this 'apocalypse' means, the total destruction of humanity on Earth, like a vast nuclear war; I believe it does not. When translated literally, the word apocalypse means; to unveil, reveal that which was hidden. Many people, like myself, believe this event is truly... the ~End of Times as We Know Them~. This means that, it will be something so spectacular; that life on Earth has never been like this before. Welcome to the Revelation.

I believe that *Revelations* has meanings, that are hidden inside of other meanings. And that Saint John, the Divine Revelator, wrote his prophesies to encode astronomical symbolism. He did it a few different ways. First, by directly stating the event; like how *Revelations* out right tells us the year of birth, for the Second Coming of Christ. And also, by encoding mathematical clues of astronomy and the calendar, into his writing. Thus, John the Revelator pinpointed the very day for the birth of the Second Coming of Christ.

St. John had achieved a very advanced understanding of the Precession of the Equinox and its entire Precessional Cosmology... i.e., the BIG PICTURE of the time table, for the Second Coming of Christ. The Revelator almost certainly came from a Christian Mystery Religion Tradition. And as such, he learned their secret knowledge. But Saint John the Divine, also had true visions of events that began 900 years after he was born. So, he had more than just secret, Precessional knowledge; he 'saw' into the future. TRUE PROPHECY.

And St. John's Christian Mystery Religion would have been one of the many Mystery Religions. Some of them were Gnostic, some were Johannites; and all of them had the same Mystery Tradition teachings about the Precession of the Equinox Cosmology. It is my belief that all the various Mystery Traditions, knew that the 'End of Times' had to happen at the Galactic Alignment... AND THEY ALL KNEW that; our own modern times; is that Galactic Alignment. Specifically, we are in the last half of the present 72-year Era. The most common way to measure this alignment is the December Solstice-sun conjunction with the Galactic Center. The Precession moves 1 degree every 72 years. 360 degrees times 72 years is 25,920 years. The closest year of the Galactic Alignment was 1998, reported from the U.S. Naval Observatory. Which means our Era is 36 years before then, and 36 years after. 1962-2034

Jesus was born in 6 BC, when the ages changed from Aries to Pisces. So why was there another specific year chosen to become 0 BC? Perhaps for this... 333 + 666 + 999 = 1998. This is from an urban legend I heard decades ago. I was told that 333 is the number for the Christ, and that 999 is the number for Armageddon. Everyone knows that 666 is the number of the Anti-Christ. So, 1998 was supposed to be the year of the Apocalypse. I do not believe this is from the Jewish numerology, and that is why I call it an urban legend. But where ever it came from, it was right for Precessional Cosmology, even if it was not originally intended to.

I believe that John, called the Divine, had learned from his Mystery School; how the very ancient masters, from long before his own time, had used a 'hidden metaphoric language' to explain Precessional Cosmology. And how they wrote these encoded lines inside the world-wide ancient mythos. And for us, there is NOW a modern book called *Hamlet's Mill,* that presents literary evidence. It reveals how this hidden language was crafted by a global endeavor of astronomy cultures, as far back as the late Neolithic Age, (5000-6000 BC?). It was used to secretly preserve knowledge about the Precession, in the oral traditions and writings; for all future generations, (p.3 HM). It started in the New Stone Age and continued through the Copper and Bronze Age and into the Iron Age. And in the Western World, all the way until the time of Jesus. These sets of metaphors were encoded into actions, and symbols, and word meanings. This was a language to tell a story. It was used by; the ancient Mystery Religions, and priesthoods of all the organized religions; including the earliest Christian Mystery Traditions. They all used this language to encode information, on a specific set of subjects; into ALL of the various world-wide sacred literatures and mythologies. Together it's a set of subjects, of which Precessional Cosmology was one topic; all leading to a BIG PICTURE. It's a pyramid, with the arrival of the Male Messiah at the apex. And the evidence shows that, JOHN, THE DIVINE REVELATOR, used this language too. So if I'm not mistaken, this makes it, one of the world's biggest and oldest conspiracy theories EVER.

We do know that during all of humanities time on Earth, there were many ancient star-gazing cultures. And many of them used the -360 degrees of a circle-measurement. Like the Babylonians and the Egyptians, as well as the Hindus, and through them the Chinese. The *Rigveda* mentions a wheel of 12 spokes and 360 pegs. It is also well known that with the great lifespans of these cultures, they had 2000 to 3000 years of tracking the heavens. Tracking every single minute, of every single hour, of every single night, ~that they could; without fail. And that, they used the circular 360-degree measurement. Remember that, Precession moves 1 degree every 72 years. That means, during 2160 years out of the 25,920 years, the stars in the sky had moved away 30 degrees, that's $1/12^{th}$ of the entire sky; an entire Zodiac sign. How can anyone think the knowledge of the Precession of the Equinox was 'first' discovered by Hipparchus, a century and a half before Christ was born in 6 BC? To me it's simply rational, that these star gazing cultures would have noticed the sky move; and it seems impossible that they did not. John Major Jenkins, Graham Hancock, and other authors have provided literary evidence.

I firmly believe that the ancient, world-wide priesthoods which tracked those stars, kept these secrets of Precessional Cosmology. It's an astronomy clock, a time table, a moving into alignment. It foretells how the Second Messiah would have to be born in the current, 72-year, Golden Era. During the whole cycle for the Precession of the Solstices and Equinoxes; there are a total of 360 eras, of 72 years each. And it is our era, which is the one during the whole cycle of 25,920 years; that clearly stops and starts this Great Year, making it a Golden Era. The secret knowledge about our current Golden Era was passed down through the generations and centuries, along with other secrets and various levels of technology.

Then around 1300 BC, according to Rosicrucian legend; the Egyptian Pharaoh, Akhenaton, did more than just enforce his monotheist god. He also gave away to the public, the priesthood's secret knowledge about the great mysteries. Akhenaton is believed to have helped start the Mystery School Tradition; so that all humans would have access to these secrets and their connected topics. One secret was the Precessional Cosmology. There were also the more flashy secrets, that the Mystery Schools promised to reveal to those people who joined them. They taught their initiates secret ways, to 'attain higher consciousness' in DEATH. The oldest Mystery School, Eleusis in Greece, started around 1200 BC; that's just one life span after Akhenaten's Era. It continued until almost 400 AD; and that's 1600 years. For 1600 times the initiates at Eleusis, outside of Athens, "SAW THE SUN AT MIDNIGHT".

The secret knowledge of Precessional Cosmology is a CLOCK, both simple and complex to understand. It has been used since the Neolithic Revolution. It shows that the point is 'now' for the stop/start, 72-year Golden Era, of what St. John the Divine Revelator wrote as the Second Coming. And it is commonly known that our December Solstice Sun is now directly aligned with the Galactic Center. But there is so much more going on, with astronomy.

It is from my own studies; my advanced knowledge of Precessional Cosmology, that I know more secrets; to what is: the GALACTIC ALIGNMENT OF THE PRECESSION OF THE SOLSTICES AND EQUINOXES. And when the whole mystery I explain is understood; it shows that we are RIGHT NOW in a ONCE IN INFINITY EVENT, during THIS ONE specific Galactic Alignment ~ out of all other Galactic Alignments. I offer evidence that this current stop/start, of the 25,920-year cycle, is very special and divinely proportioned; more than any other 'GREAT YEAR'; that has ever been, or will be.

....

PART I – The Visions of Saint John.

Chapter 1: *Revelations XII*

And now I will proceed with the prophecies of Saint John the Divine.

Revelations has cryptic passages, that have had their meaning guessed upon by many people, all through its 1900+ years of existence. And for most, what stands out more than the rest; are the very first words of *Revelations 12*; which speaks about the first appearance of the Dragon. Specifically, in passages 12:1-6, and then again in 12:14.

> "Behold, great portents appeared in the heavens! A Woman robed in the sun, with the moon under her feet, and a Crown of twelve stars upon her head. She was pregnant and cried out in pain, about to give birth. Then another portent appeared in heaven: an enormous red Dragon with seven heads, and ten horns, and seven crowns on its heads. Its tail swept a third of the stars from the sky, and flung them to Earth. The Dragon stood in front of the Woman about to give birth, so that it might devour her Child, when it was born. She gave birth to a son, a Male Child, who 'will rule all the nations with a Rod of Iron.' And her Child was taken up to God and to his throne. The Woman fled into the wilderness, to a place prepared for her by God, where she was taken care of for 1,260 days."...**And then**... "And to the Woman was given two wings of a great Eagle, that she would fly into the wilderness."

Listed below are the clues I have uncovered a meaning for…

1) PORTENTS IN THE HEAVENS
2) THE DRAGON; 7 HEADS, 10 HORNS, 7 CROWNS
3) THE DRAGON'S TAIL
4) 1/3 OF THE STARS FALLING TO EARTH
5) THE WOMAN WITH THE SUN AND MOON
6) THE CROWN
7) 12 STARS
8) THE MAN-CHILD BABY
9) THE ROD OF IRON
10) 1260 – DAYS
11) THE EAGLE'S WINGS

PORTENTS IN THE HEAVENS... Taken literally a 'portent' is a 'sign', a way of predicting or divining the future; and it also means a way to understand hidden things. And there is the other clue-word, 'heavens', which also translates to the 'sky'. For me, it literally means the starry sphere of outer space. So, the 1st clue to understand the vision of the DRAGON and also the BEAST, is that they refer to the starry sphere. And obviously; it is events that apply to the future, from when it was written. *Revelations* was always understood to be about the 'End of Times as We Know Them'... About hidden things that will be revealed, and the presence of God Almighty on the Earth.

....

Chapter 2: Precessional Astronomy Basics

The ancients used astronomy to pinpoint the era of the Apocalypse. They did this by understanding which areas in the starry sphere are important, in Precessional Astronomy. They knew where those are, and WHY those plotted points in the heavens show an undisputable CLOCK; that pinpoints when the 72-year Golden Era, of the 'end of times', would start and stop the Great Year.

And so, very much from *Revelations 12* is about the Galactic Alignment of the Precession of the Equinox, that 'WE' are 'CURRENTLY' in. And the best way to have a secret system for mapping the starry sphere is obviously with constellations, and to encode the clues into the stories about them to preserve the map. These are the 'Mythos' that have lasted through millennia, lasting from all over the world. So many of the mythos talk about the new ages and eras that their cultures came from and are passing to, the so-called WORLD AGE MYTHOS.

....

Chapter 3: The Celestial & the Ecliptic

There are 2 planes of gravity created by the Earth. These 2 planes each surround their own axis. And of course, each axis has its north and south pole.

The Celestial Axis, with its north and south poles in the sky, is where the stars appear to revolve around in the 24 hours of the day. But as everyone knows, this is an illusion caused by the Earth spinning on its Celestial Axis. The Ecliptic Plane is the annual orbit of our planet around the sun. Astronomy calls this plane simply, the Ecliptic. However, I call it the Ecliptic Equator, to be specific. As it is perpendicular to the Ecliptic Axis with its North and South Ecliptic Poles, like the Celestial Axis is with its Equator. The Zodiacs are the constellations that this Ecliptic Equator crosses through at the two Equinox points. The spinning Celestial Axis is titled roughly 23 and 1/3 degrees away from this Ecliptic Axis.

NORTH ECLIPTIC POLE   NORTH CELESTIAL POLE

The Celestial and Ecliptic are tied together. As the Celestial and the Ecliptic Axis' have their perpendicular planes, the Celestial and Ecliptic Equators; each Equator is permanently joined to its specific Axis. And also, the Celestial and Ecliptic planes are tied to each other in two ways.

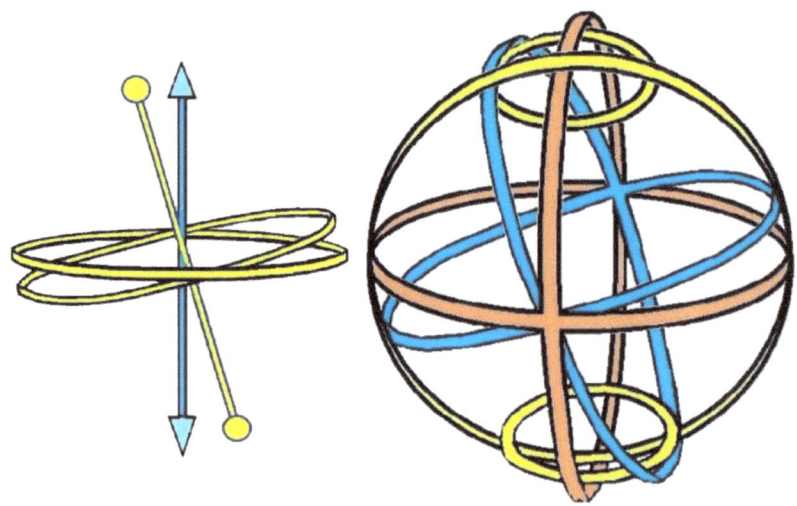

First, with the 23 and 1/3 degree tilt/separation between the daily spin and the yearly orbit; it bisects the two planes of gravity in the middle. We call it the 'Equinox'. It's the line where the Celestial and the Ecliptic Equators meet. It goes from one Equinox; through the center of the earth, where the Celestial and Ecliptic axis meet; and continues to the other Equinox. When astronomers represent all of this, we use rings, and there's more than just the 2 equators. There are 5 thin rings, all of the same size, all joined together other; to make a sphere. These rings are called 'colures'. The drawing of them all together in their mechanical device is called an ARMILLARY. And it has only a few essential parts. It is made of 2 Axis' and 5 thin rings, and 2 much smaller rings. This one type we are talking about is the Ecliptic Armillary. The other Armillary, with the tropics of Cancer and Capricorn, is the Celestial Armillary. These are types of the Stellar Mechanism.

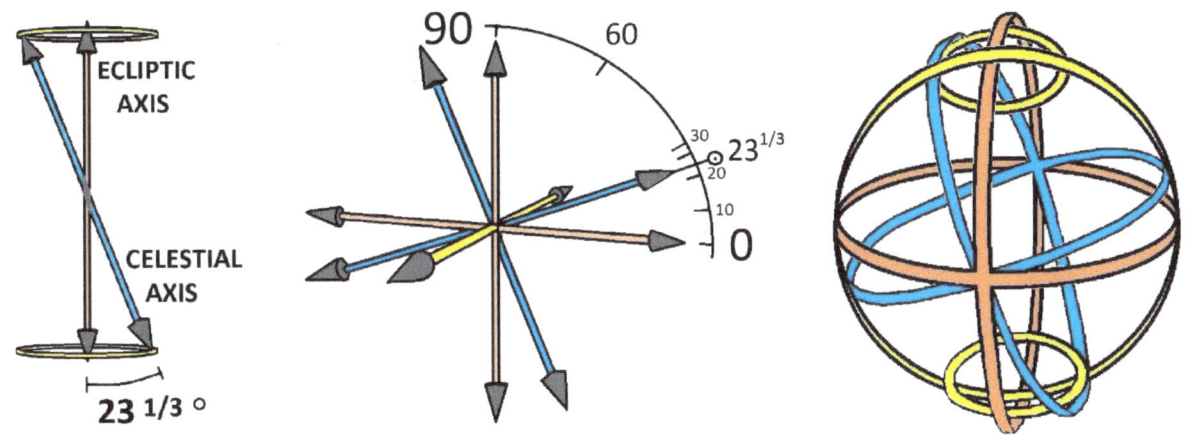

In the Ecliptic Armillary, with its 5 rings; 4 of them are the 2 Equators, and the 2 Equinox Colures (because there is one Equinox Colure for each; the Celestial, and the Ecliptic). Each of the 2 Equinox Colures encircles and connects together the specific north and south poles of their Axis', with the 2 shared Equinox spots, where the Ecliptic and Celestial Equator's meet. Thus, each of the 2 Equinox points, connects all 4 rings; both the 2 Equators and 2 Equinox colures, to a single point.

And finally, there is 1 Solstice Colure, that is shared between all of the other 4 rings. It binds the other 4 colures together at the widest points between both Equator rings and both Equinox rings. This creates a 4-point, equal spacing along the Equators (between the 2 Solstice points and the 2 Equinox points). And so, it makes the 4 directions; North, South, East, and West. Together they create the points of the December Solstice, September Equinox, June Solstice, and March Equinox.

Chapter 4:
The Celestial Axis swivels around the Ecliptic

There are some key concepts that explain the Precession of the Equinox and the Galactic Alignment. For those not familiar with this, it might be good to read it a second time... It's easier to pick things up the second time around.

Relatively speaking, the Earth spins on its axis superfast every day... and it ALSO wobbles, or swivels; really super-duper, slow, in a small circle. That slow swivel is the Precession of the Equinox. The Celestial Axis pivots around in the center, where it meets the unmoving Ecliptic Axis. Both the North and South Pole of the spinning axis make a small circle pathway.

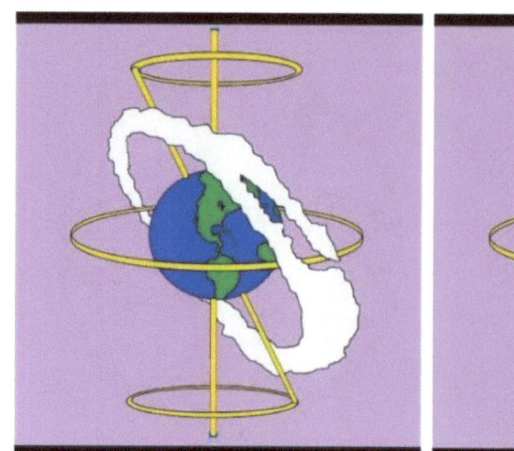 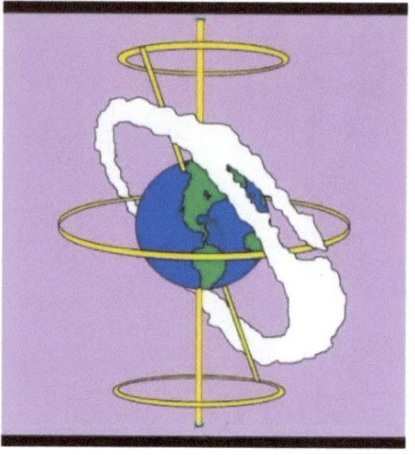 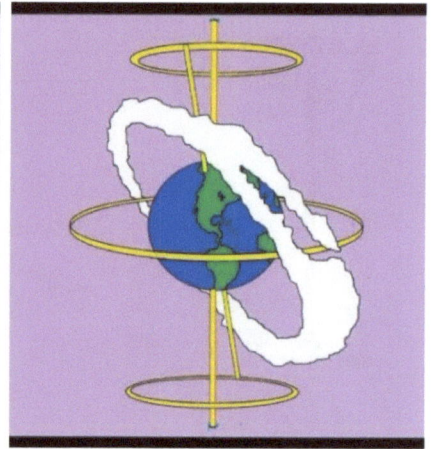

And these upper and lower circles are roughly 23 and 1/3 degrees, away from the straight up and down Ecliptic Axis. Just like the polar regions of the Arctic and Antarctic Circles are 23 & 1/3 degrees away from the Celestial Axis (where the days of complete darkness start). What this means is that, the Celestial-daily-spinning Axis; swivels in a circle taking thousands of years to complete. These 2 tracks, that the North and South Celestial Poles move around, are called the 'Circles of Precession'.

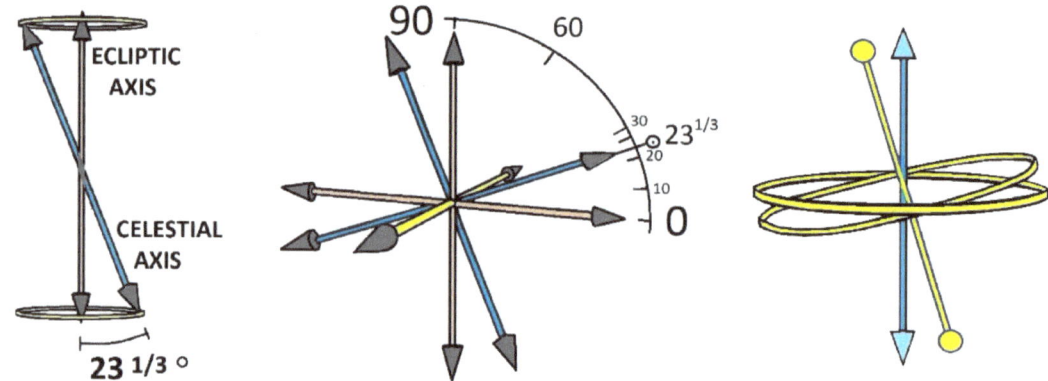

And just as the North and South Celestial Poles move along on their Circles of Precession; so to do both Equinox points and both Solstice points, move in a circle. And it's all tied together. The whole movement of the Stellar Mechanism, the Ecliptic Armillary, ever so slowly rotates the 4 points of the Solstices and Equinoxes. And it moves 1 degree every 72 years.

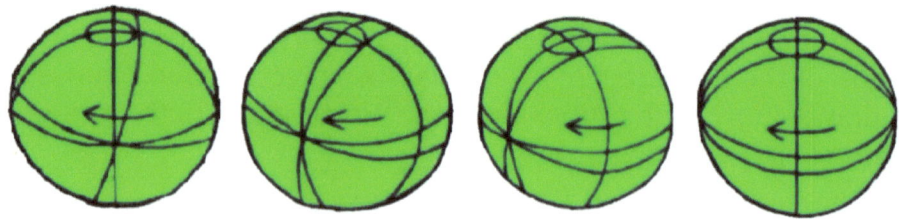

UNDERSTANDING REVELATIONS BY ASTRONOMY

Chapter 5: The EYE of God, Heaven, Horus, Sauron

The 2 Circles of Precession are much smaller than the 5 Colures. All together they make the EYE, which is the full Ecliptic Armillary. The Eye became a major symbol for knowledge of astronomy. The concept expanded out, and so now the Eye also refers to a more advanced Armillary... As a PRECESSIONAL CLOCK. That's when the Ecliptic Armillary also includes the Galactic Equator... Altogether, it creates the 'Eye and the Pyramid'. And later in this book, the 'Pyramid' is explained also.

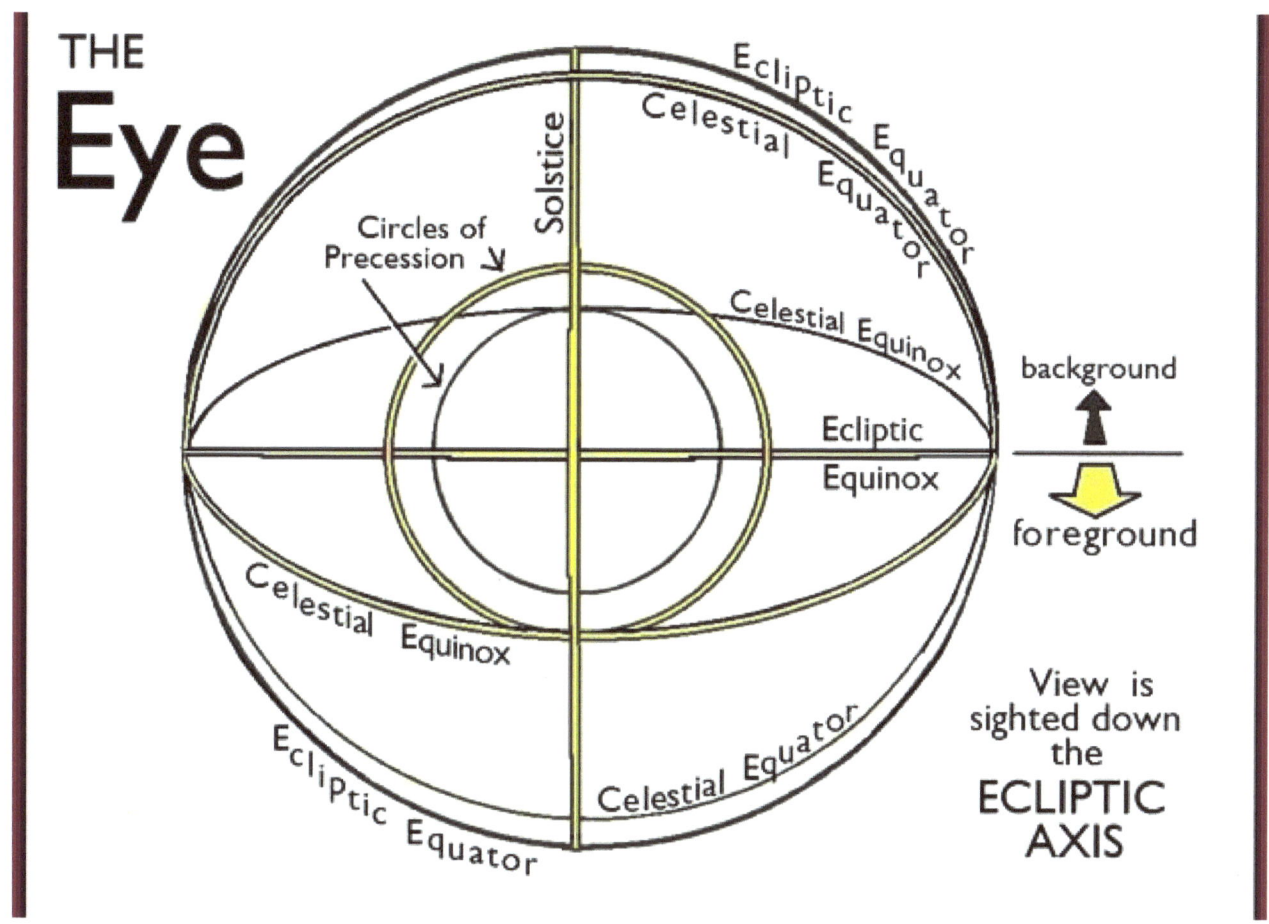

To see the Ecliptic Armillary as an Eye it must be sighted straight down the Ecliptic Axis. And from this view, the 'EYE' has an oval lid tight around an iris and a very dilated pupil. It also has the full circle around it, suggesting the sphere of an eyeball, with a bull's eye cross. Pictures are worth 1000 words.

....

Chapter 6: The EYE moves

What Precession means FOR THE CALENDAR, is that; -1) those calendar dates of the Solstices and Equinoxes always stays the same. These dates are roughly the

21st of Dec., Sept., Jun., and Mar. - 2) And on Earth, it looks as if the stars of the night time sky slowly revolve around, infinitely slow with Precession; passing through those dates. But really, it's the dates of the calendar that ever so slowly spin around the fixed starry sphere. And they move 1 degree every 72 years, for the total of 25,920 years.

And now with the slow turn of Precession, seen from the hemisphere view sited down the Ecliptic Axis; we add the last important feature. The Galactic Equator. That's how the measure for the stop/start is determined.

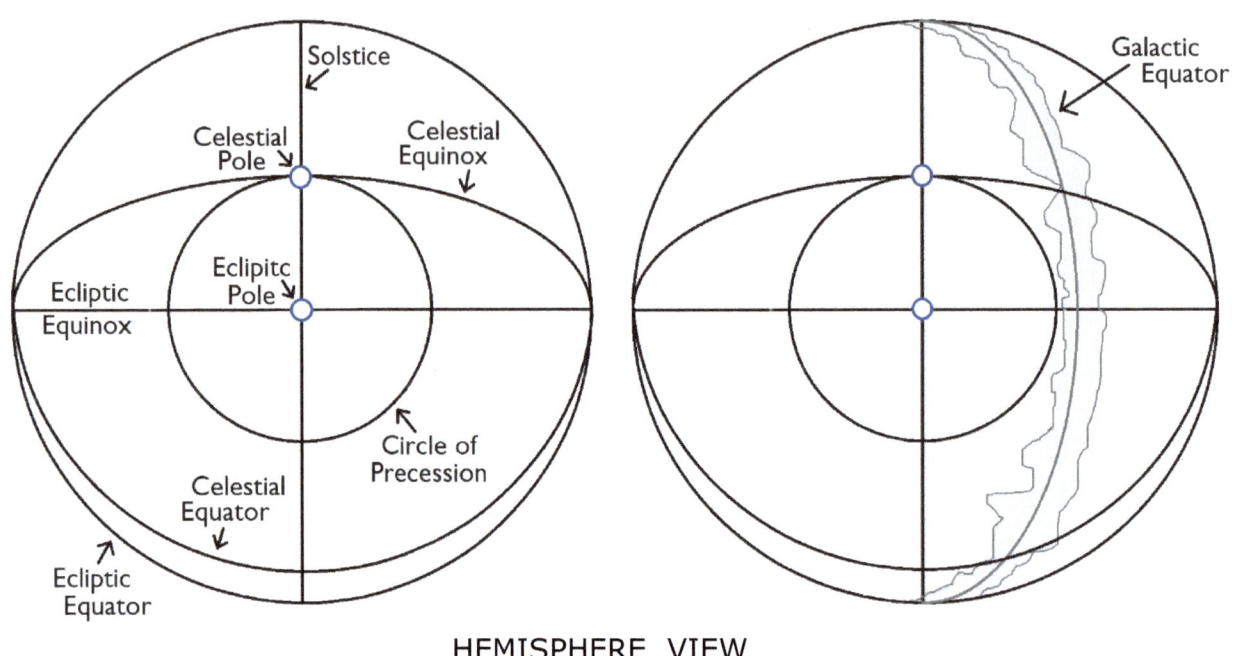

HEMISPHERE VIEW

With the Galactic Equator in the diagram, it shows that the whole mechanism spins around in a slow circle, inside the Ecliptic Equator. So that Celestial Pole for the Ecliptic Armillary of rings, spins around the unmoving ring of the Galactic Equator.

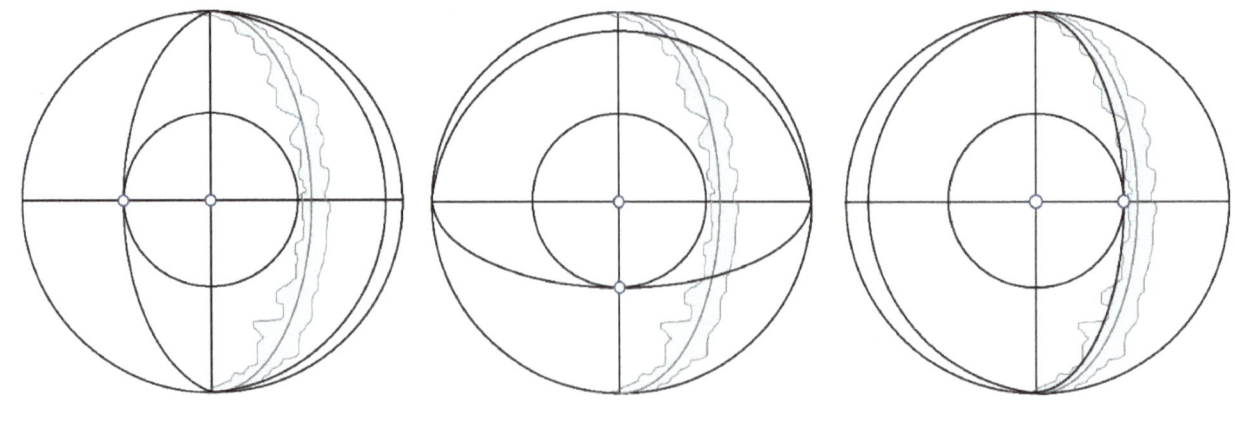

UNDERSTANDING REVELATIONS BY ASTRONOMY

Chapter 7: The Galactic Alignment

The ancients assigned a specific number of years for the slow moving Precession of the Equinox, it is 25,920 years. Many scholars have realized that this is equally divided by the number 108. Thus, 25,920 divided by 108 is 240. But to my knowledge, until now, no one has ever published the reason why 108 is so important. But that is a later chapter.

The Galactic Alignment is like one very, very, long-time-coming New Year's Day. But actually, it's 1 degree out of 360, called a 'Golden Era', so it's 1 period of 72 years, out of 25,920 years. And the math is simple; there are 360 degrees in a circle. And 72 times 360 is 25,920… And what's going on in our lifetime, is that WE ARE IN THAT SPECIAL 'GOLDEN ERA' RIGHT NOW. BUT, the question is… in the heavens, what is happening to pinpoint our 1 period of 72 years, out of all the 360 periods of 72 years? The answer starts with this…. The Earth sits in the Galaxy, and one tiny direction of the starry sphere points to the center of the Galaxy, called the Galactic Center.

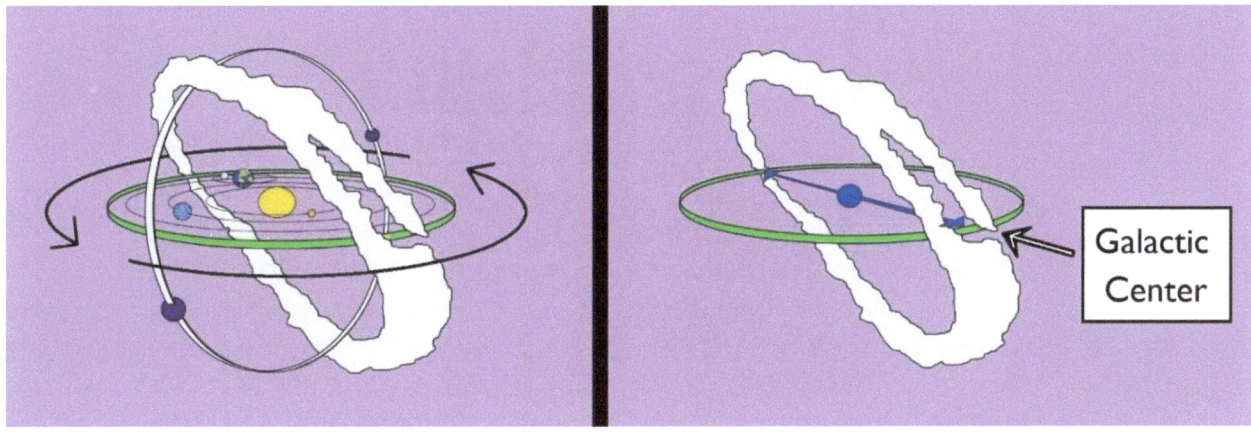

Remember, the stars only 'appears' to move through the calendar dates. The Galactic Center 'appears' to move through the each of the 2 solstice dates and the 2 equinox dates, going through the entire calendar year. However, it's the calendar with the Solstices and Equinoxes that actually moves; not the stars, or the Galactic Center.

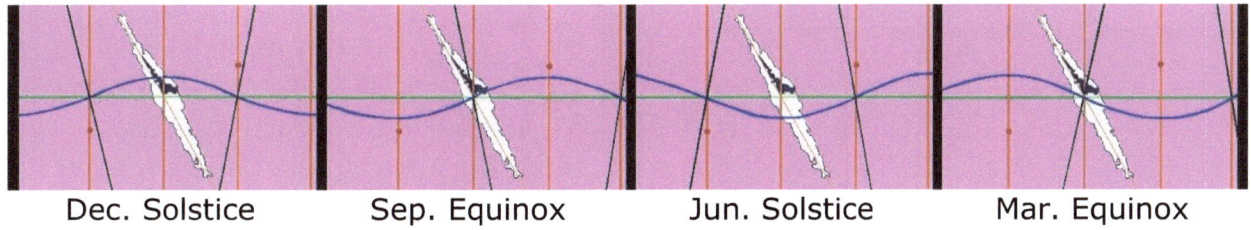

Dec. Solstice   Sep. Equinox   Jun. Solstice   Mar. Equinox

UNDERSTANDING REVELATIONS BY ASTRONOMY

It's all connected together. Since our Galactic Center is aligned with the December Solstice; the Ecliptic Equinox Colure, aligns 45 degrees away with the North and South Galactic Poles. Later; as the each Equinox point aligns with the Galactic Center; the Solstice Colure aligns with the North and South Galactic Poles.

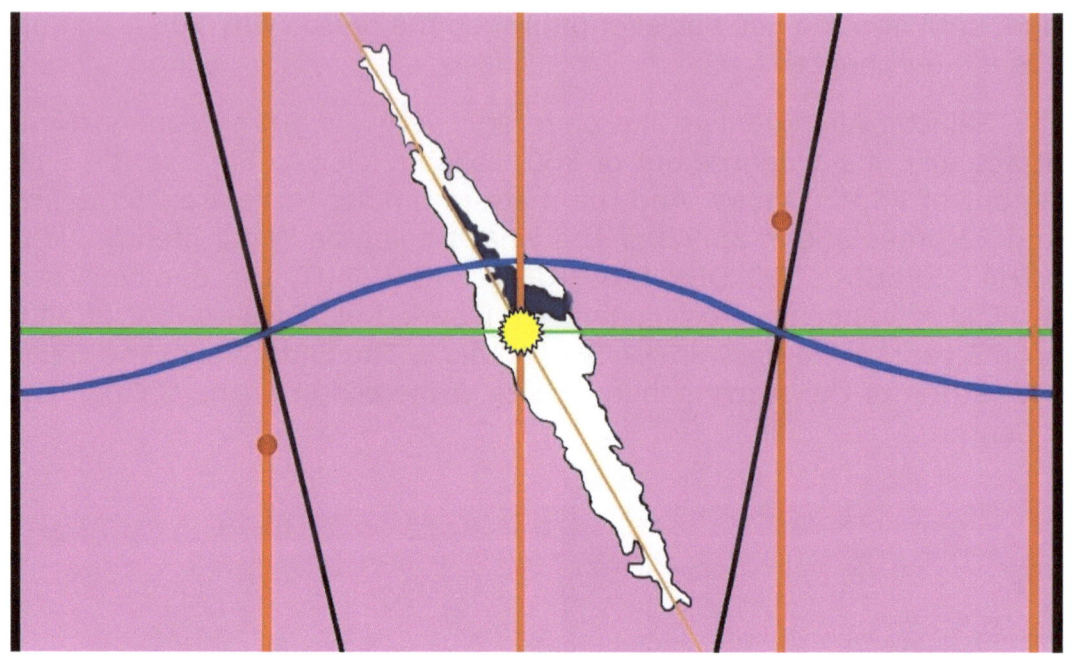

DECEMBER SOLSTICE & GALACTIC CENTER CONJUNCTION

What this Galactic Alignment means on a practical level is this... Right now, in this 72ish-year period; the Galactic Center is directly behind the sun as it crosses through the sky every December 21st. And 72 years in the future, the Galactic Center will be behind the sun on Dec. 20th. And in another 72 years after that, the Galactic Center will be directly behind the sun on Dec. 19th. You see how the dates precede backwards and do NOT proceed forward. This is why it's called the Precession of the Equinoxes and Solstices, because the dates precede backwards through the starry sphere.

There's a lot more than just that above, about the Galactic Alignment and the Precession; and for those of you who are Archeo-Astronomers and Precessional Cosmologists like me, perhaps YOU already know all that I have stated so far. And as the book goes on you may see a little more of what you already know, if you have read *Hamlet's Mill* or *Maya-Cosmogenesis 2012*.

Chapter 8: The Year of the Messiah's Birth

Here's a very simple clue explained from *Revelations 12*. Remember, what we are talking about is NOT just the entire 72 years of the Galactic Alignment. But specifically, THIS IS ABOUT THE ARRIVAL FOR THE SECOND COMING OF CHRIST, during the Galactic Alignment. It's the BIRTH of the MAN-CHILD who holds the ROD OF IRON. And that birth is just 1 year out of the whole 72 years of this 'GOLEDN ERA'. So, what's the year? The clue is this: "the Dragon casts its tail through the heavens and one third of the stars fall from the heavens to the earth." This is the *Revelations* of Saint John the Divine… And the vision of Saint John the Revelator was given to him almost 1900 years ago. So, while many of the clues here can be put down to very rare knowledge about long observed phenomenon… THIS ONE, IS A TRUE PROPHECY. Rather than present all the information first, and then give the year of the male Messiah's birth; I will give the year, and then explain why it is that date. THE MALE MESSIAH, THE 2ND COMING OF CHRIST; WAS BORN IN 1966. Because this was the only year so far, in our 72-year period of the Galactic Alignment, when 1/3 of the stars fell from the heavens. Indeed, it's the only year ever in recorded history, when 1/3 of the stars fell.

On 17 November 1966, one third of the stars fell from the heavens. It was the greatest Leonid Meteor Shower superstorm ever recorded. And they have all been recorded. The Leonid Meteor Shower happens every year, but once every 33 years there is a superstorm. Lots and lots of shooting stars! And it just so happens that it occurs at the turn of each century; at '00, then '33, and finally '66. This is one meaning for the repeated numerology of 'one third' from *Revelations 12*. It happens every one third of a century, tied to the turning of the century.

This was a true prophecy! Because the Leonid Meteor Shower started in the year 902 AD. So, Saint John the Divine made his prophecy about 800 years in the future. In the 1400's the Leonid Shower superstorm stopped for 100 years, before it continued again. And then recently, in 1866 the Leonid superstorm was spectacular. It was the best ever seen since, before that year. And then, the superstorm was gone for 100 years. BUT, in 1966 it came back. The biggest Leonid firestorm ever seen, and maybe that ever will be seen. At its peak, OVER 40-60 SHOOTING STARS… EVERY SECOND… FELL TO EARTH.

The 1999 meteor storm was very unspectacular. And the 2033 firestorm in the future, happens on the very cusp at the end of the 72-year period; since the midpoint was 1998, the last year is 2034. And in my opinion, for what needs to happen to the Male Messiah, to kick in his phenomenal cosmic powers; a one-year old baby would not be up to the task. And yet, who knows? But my best guess is 1966. AND THIS MEANS THAT THE MALE MESSIAH IS ALREADY ALIVE ON EARTH!!!

And we can pinpoint the exact day that this male messiah was born as well. All from the clues in *Revelations*. But that's for later.

Chapter 9: The Crown of 12 Stars

Now for the vision of the Crown with 12 stars. Again, the word stars tells us that it's about... stars. And this clue is very easily understood with a little research. Many of the earliest Christians, as well as the hundreds of millions of living Eastern Christians, believe that the Father, Son, and Holy Spirit are not the same entity. Yes, they are all divine, but they are 3 separate beings. There is God the Father; who made creation, but who is not inside creation; He is transcendent. And there is God the Son; Jesus the human, who opened heaven for all humanity. And instead of just the Holy Spirit, SHE is God the Mother, called Holy Sophia, Wisdom; and She is the presence of God in all of creation. And she wears the crown of twelve stars. And what is also known from that far off time of 2000 years ago, is that the 'twelve stars' refer to the 12 Zodiac Constellations of the Ecliptic.

The Precession of the Solstices and the Equinoxes, takes 25,920 years to move though the entire Zodiac series; in one great revolution called the 'Great Year'. This means that, for instance, the December Solstice takes 2160 years to move through each Zodiac 'SIGN'. 2160 times 12 equals 25,920... But the Signs borders are not the exact Constellations. Each of the Zodiac Constellations are a little shorter here, and a little longer there, from one another. The 12 Signs are equal spaced. And technically, there are 13 Zodiac Constellations that touch the Ecliptic Equator. The 13th constellation is called OPHIUCHUS, and for 'some' reason, even though more of it touches the Ecliptic Equator, than its neighbor the 'Scorpion'; it's not considered to be in the Zodiac. The Constellation Ophiuchus also comes to play again later.

So, the Zodiac Signs are 12 evenly spaced sections of 30 degrees, pie-divisions of the circle. Now, getting back to the CROWN OF 12 STARS; since it refers to the 12 Zodiacs Signs / Constellations; it refers to the 'Great Year' of the Precession of the Solstices and Equinoxes, and its 25,920 years. The Crown of 12 stars is both the Great Year, and the Solar Year

One part of St. John's *Revelations* is known to be connected to the Zodiac, along with the vision of *Ezekiel*. They both saw wheels in the sky with eyes all over them, and they both describe almost the same 4 Angels. The Bull, the Lion, the Human, and the Eagle. And these are well known to represent the 4 constellations of; Taurus, Leo, Aquarius and the Eagle. In *Revelations,* each 1 represents an Angel. In Ezekiel's vision the 4, are joined together to make each 1 of the 4 Angles; so each Angel has 4 FACES, made from the 4 constellations. 3 of these are zodiacs. And then there's the Eagle Constellation, Aquila. But, why not the Scorpion Constellation? We shall see!

The next part of the vision is THE WOMAN ROBED IN THE SUN WITH THE MOON AT HER FEET. And of course, she's the one with the CROWN OF 12 STARS ON HER HEAD. And this is, in my opinion, a reference to Holy Sophia, God the Mother, or if you will, the Goddess. As a constellation, it seems that the likely choice would be Virgo the Angel. But is it the only choice? As I shall explain later there is another, a hidden candidate. And as for, "robed in the sun with the moon at her feet," many people take that to mean an eclipse. And that's a very good guess, yet it could also mean simply that they pass each other in the sky just not, in a-line-ment, with the Earth. But for what I am about to explain, I do not think the sun and the moon play a part in it.

....

Chapter 10: The Galactic Center

And what I have been calling the Galactic Center is not exactly the 'true' center of the galaxy; it is the 'historic' Galactic Center. But it is very, very close to it. What's happening is this, as the Precession of the Solstices and Equinoxes slowly spins around in its 25,920 years, it is NOT marked off against the 'true' Galactic Center. Instead, it is measured against where the Galactic Equator crosses the Ecliptic Equator. And this is NEAR that huge black hole at the Center of the Galaxy.

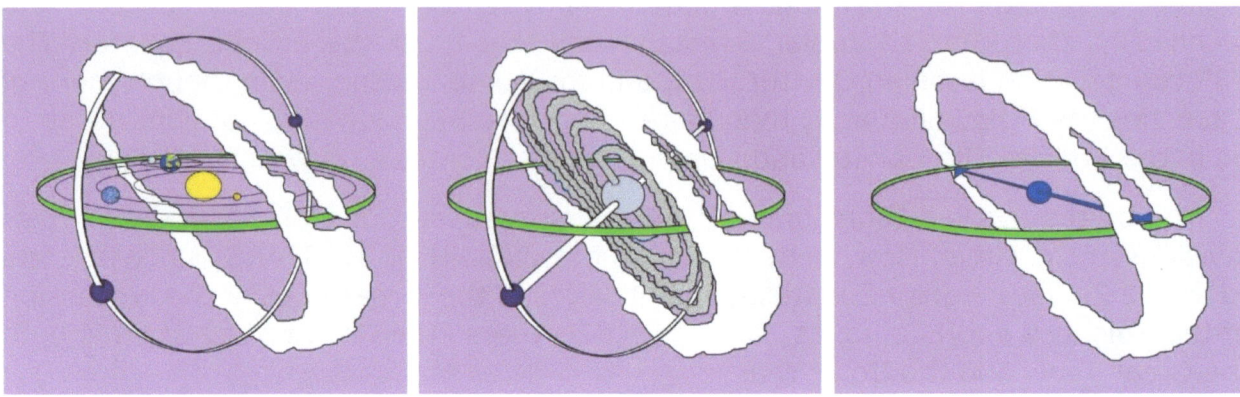

This crossing of the Ecliptic and Galactic Equators is historically referred to as the Galactic Center; even though it's not exactly correct. Since both the Galactic and Ecliptic Equators are rings that intersect; that means they connect at two opposite ends, like the Earth's Equinox. The historic Galactic Center is at one end; and the other end is called the Galactic Anti-Center.

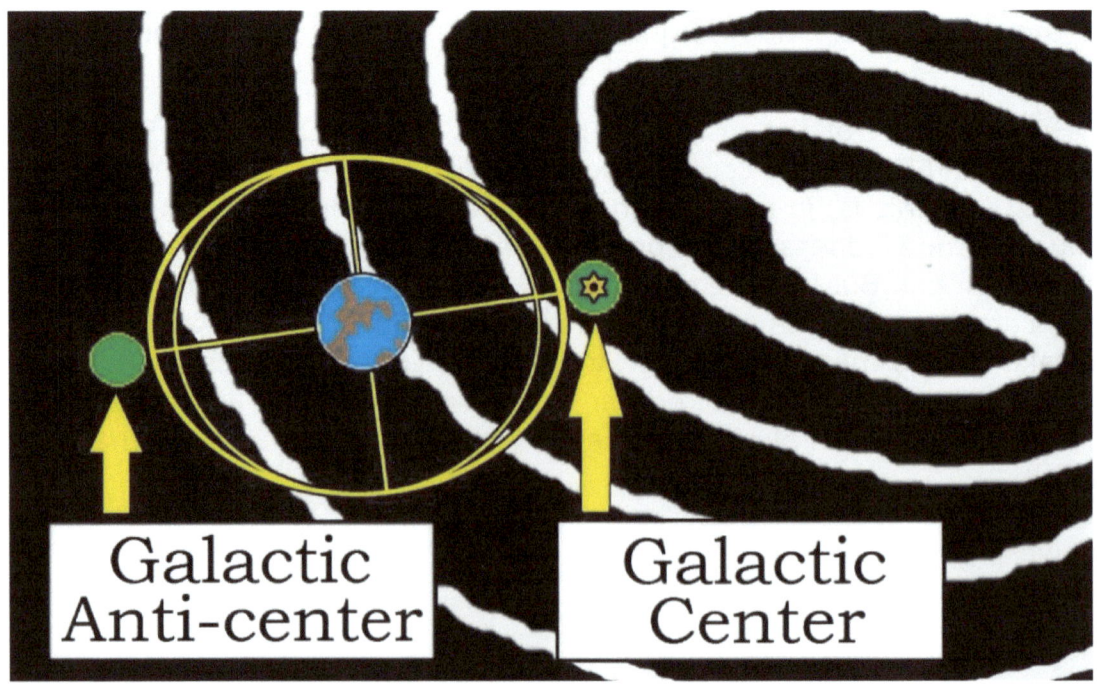

UNDERSTANDING REVELATIONS BY ASTRONOMY

Chapter 11: Dividing the 'Great Year'

The Signs of the Zodiac, came after the Constellations; the Signs are equal, 30 degrees each. The 'Signs' are used for astrology, and for the 12 equal Zodiac divisions of the 'Great Year'. Remember the song, *Age of Aquarius*? That's the World Age Zodiac Sign, and it takes 2160 years to move through each sign. This is one way the Old World divided the Great Year.

However, before the Great Year of the 12 evenly spaced Signs of the Zodiac; the Constellations along the Ecliptic Equator were used for the Zodiac World Ages. And to count the passage of time during the cycle of Precession; the stars 'seen' to be moving ever so slowly were measured off against the horizon on the March Equinox, at just before Dawn. Each Zodiac Constellation would ever so slowly appear more and more in the predawn sky.

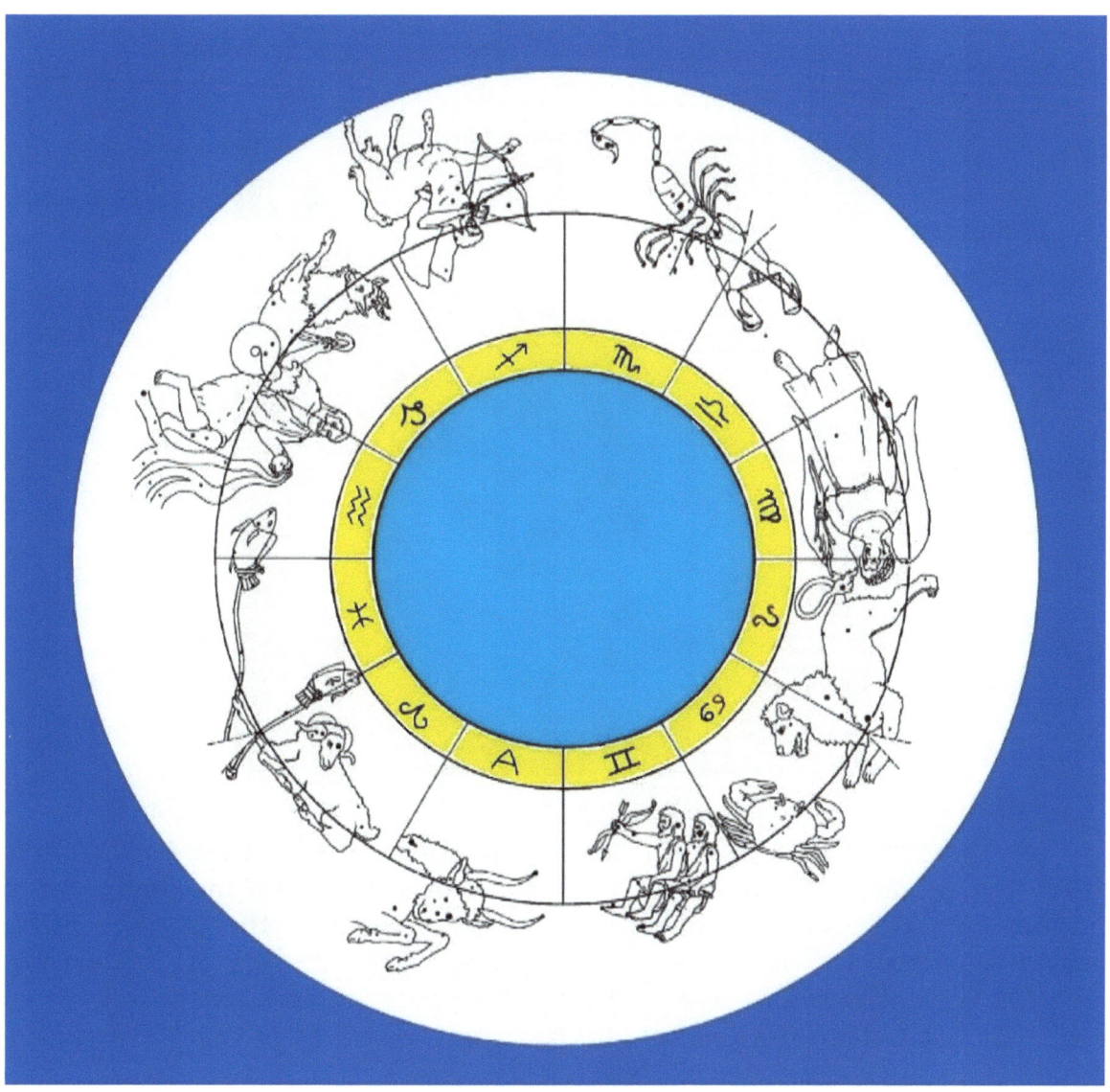

When the first star of a new constellation was seen in the brightening dark of the predawn sky it would herald the new Age of that constellation. The idea of the - new age is what gave the 'New Age Culture' its name. Then over the next 1000 to 2000 years, that constellation would slowly show itself, more and more at dawn on the March equinox; 1 degree every 72 years. And with Precession, the stars and constellations were seen to move backwards from the way the 12 zodiacs appeared in normal order during the solar year.

And also; the Great Year is divided into 5 equal parts as the 'Ages of Metal'. But that's because FIRST the Great Year is divided into 4 equal parts from the Solstices and Equinoxes; and a 5-part division that is placed over the 4 creates a dynamic life cycle. When each of the 4 Solstices and Equinoxes aligns with the Galactic Center in its turn, it is a GOLDEN ERA. So, during the entire cycle of Precession there are 4 times that we are in a Golden Era; with each one lasting 72 years. And this is why the full name is, the Precession of the Solstices and Equinoxes. But again, it's the December 21 Golden Era that is the stop and start point of the entire 'Great Year'.

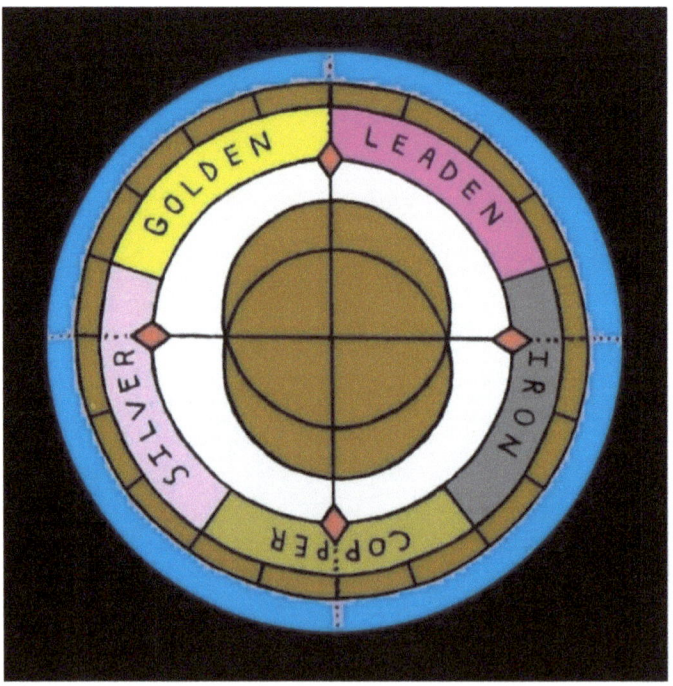

The dynamic Great Year for the Metal Ages starts with the Dec. Solstice between the Leaden 5th and Golden 1st of the Metal Ages. It's the death and birth point. The 2nd Golden Era happens in the first quarter of the 2nd Age of Metal, powering the youth for the World Age of Silver. In the 3rd Age of Copper, the 3rd Golden Era is right in the middle of the Age; which is perfect for the midlife cycle. The 4th and last Golden Era is after the last three quarters of the 4th Metal Age of Iron, powering the last quarter of life.

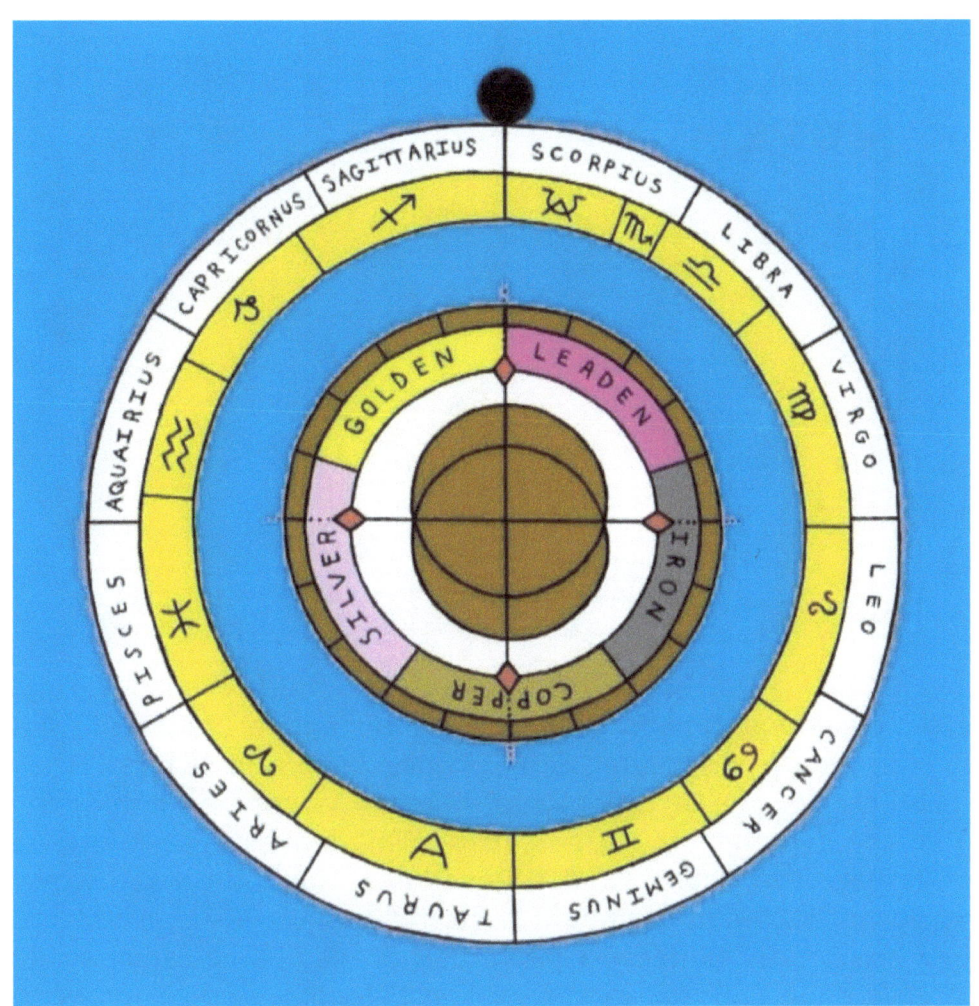
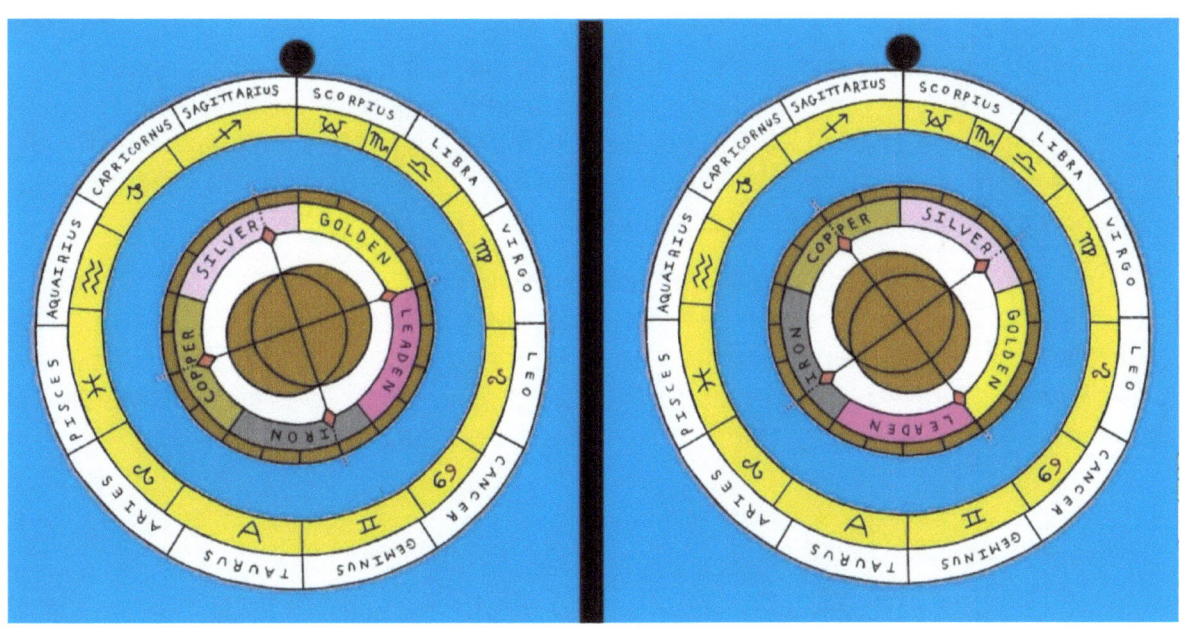

UNDERSTANDING REVELATIONS BY ASTRONOMY

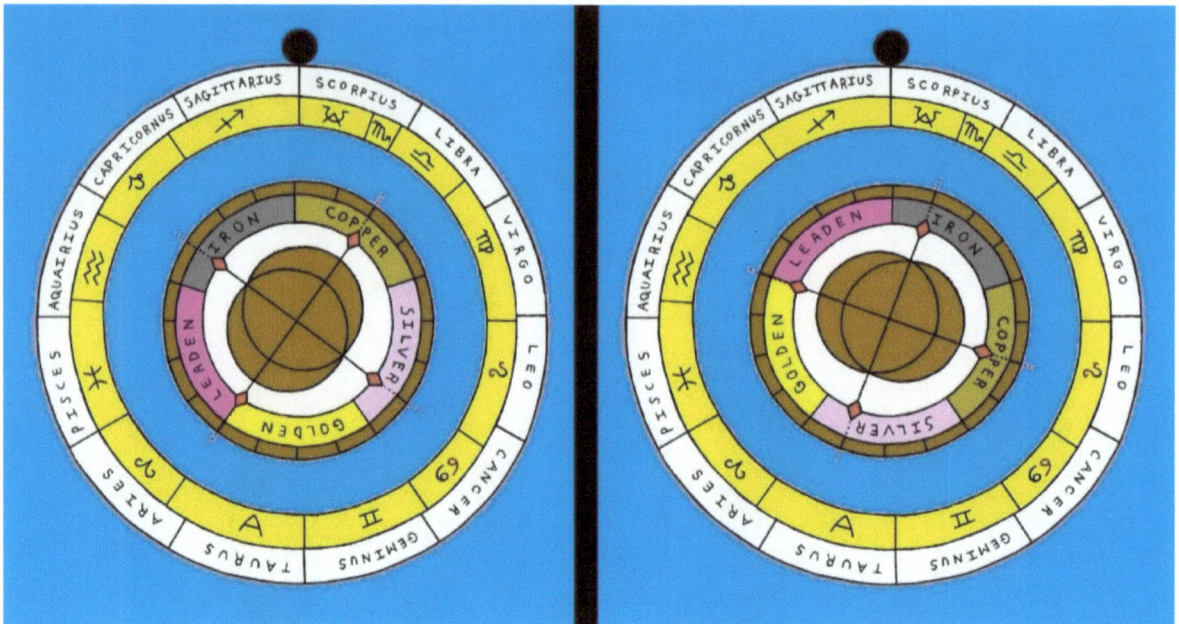

Different traditions sometimes use bronze not copper, for the 3rd Age of Metal. Or clay-and-iron for the last Metal Age. I chose Lead and Copper. They are pure elements. Lead for the age of bullets and nukes. The important thing is that the 5 into 4 created the dynamic life cycle concept. This makes a 20-fold division of the Great Year. Perhaps the Mayans and Toltecs knew this, and that's why one of the reasons they used a 20-number count system. Perhaps it's one reason why their predecessors, the Ancient Olmecs, created their 20-day name calendar. 20 days and 13 numbers create the 260-day sacred Tzolkin Calendar. And as we will see later in this book, both the Old World and New World traditions where aware of the importance of 260 days.

Now back to the Constellations of the Zodiac. Each Zodiac Constellation on the Ecliptic Equator has another Constellation directly opposite. But since they have different sizes, there are only 2 breaks between constellations that are exactly half a circle apart from each other, 180 degrees. Those 2 are the spots of the Galactic Center and the Galactic Anti-Center.

The two Zodiac constellations on either side of the Galactic Center are Sagittarius and Scorpio. And the two constellations on the other side at the Galactic Anti-Center are Gemini and Taurus. So obviously these two places in the starry sphere were recognized and considered important enough to stand out as the only Zodiac constellation divisions that are 180 degrees apart.

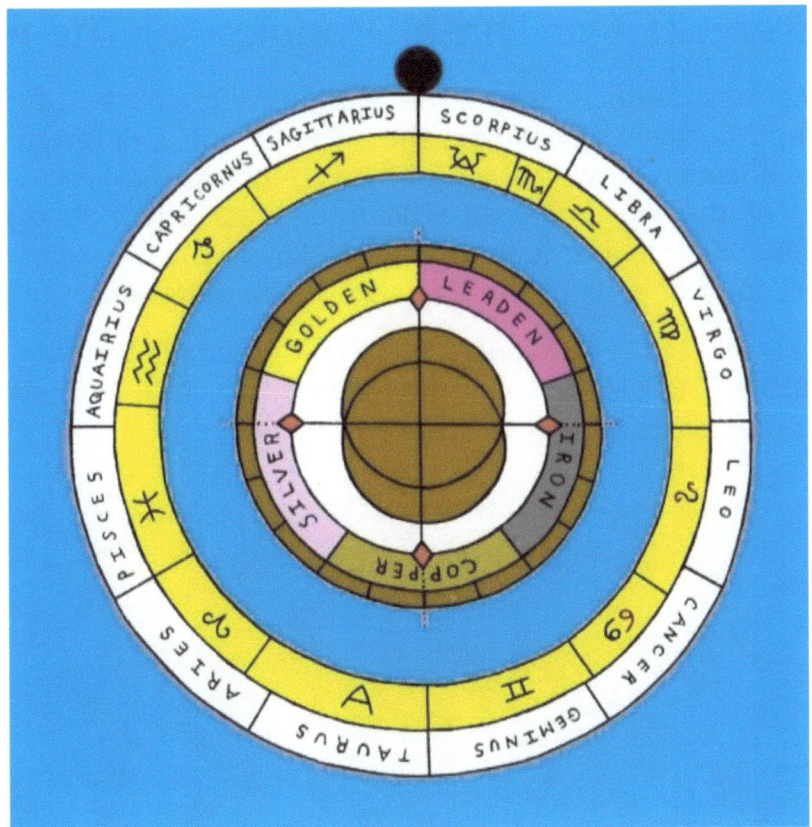

Also, it's at the Galactic Center and the Galactic Anti-Center where the divisions between the 'signs' line up with the divisions of the 12 Constellations. The only other division of Constellations that lines up with its sign division, is between Aries and Pisces; where the Zodiac is said to stop and start. Aries starts at March Equinox. In ancient times March Equinox started the Calendar year, as the 1st day of the month. So, the 1st 10 days of old March are now the last 10 days of new March. This March Equinox is the 1, of the 4 Solstices and Equinoxes, that refers to the Precession of the 'Equinox'; for which the 12 Zodiac World Ages were seen in the predawn sky.

So, the 5 World Ages of Metal are tied to the moving Solstices and Equinoxes. And the 12 World Ages of the Zodiac are tied to the unmoving starry sphere. Both systems where used. This is one reason that the Precession was called the Cosmic Mill. A mill has two stones. One moves slowly and the other stone stays still. On one level this means the moving Celestial stone; grinding away slowly on top of the unmoving Ecliptic Stone. On another level... It's the moving stone of the 5 Metal Ages with the Solstices and Equinoxes; grinding away slowly against the unmoving 12 Zodiac World Ages of the Starry Sphere.

………

Chapter 12: The Three Dragons, 1) The Hydra

One of the most famous parts of *Revelations* is of course... the DRAGON, also called the Serpent. But as we shall see, the ancients had 3 different Serpents in the heavens. So the question is, which one of them refers to the Precession of Solstices and Equinoxes. And the answer is... all of them do. Each one in its own way points to a greater understanding of not just the Precession of the Solstices and Equinoxes, but also, they all point directly to the Galactic Alignment itself.

The first serpent in the heavens is the HYDRA, and it's the longest constellation in the starry sphere. There were some really great art works in the Middle Ages and later, that show the constellations of the starry sphere.

Some of the charts, show that the Hydra constellation loosely mirrors the Milky Way. And when we observe a star chart of Milky Way on what's called a cylinder projection, like the constellation chart below, we can see that the HYDRA constellation definitely parallels the Milky Way, and that they both share similar dips and rises.

And now we get to, how this constellation points to the Galactic Alignment, which is the stop and start point of this Great Year. We know that the Ecliptic Equator is almost completely unmoving, but the Celestial Equator itself, during Precession, wobbles like a coin that was spinning and now is slowing down and about to rest on a table. The wobble is a full 46 and 2/3 degrees wide. And the upper and lower ends of the wobble movement are the tropics of Cancer and Capricorn. Both are 23 & 1/3 degrees from the Equator.

Right now, during this Galactic Alignment... As the Celestial Equator has wobbled another one full turn during the 'Great Year', the Celestial Equator cuts the head off the HYDRA. This of course is one of the 12 labors of Heracles; and the 12 labors are also known to represent the 12 Zodiacs in their own encoded way.

Ogygia means 'primordial' and is an ancient name for the South Ecliptic Pole. It is clearly marked by the large Magellanic Cloud. It was the Golden Island, With the Golden Cave, with the Golden Bed, with the Golden Age God sleeping. The Once and Future King; Saturn, the God of the Ecliptic Axis.

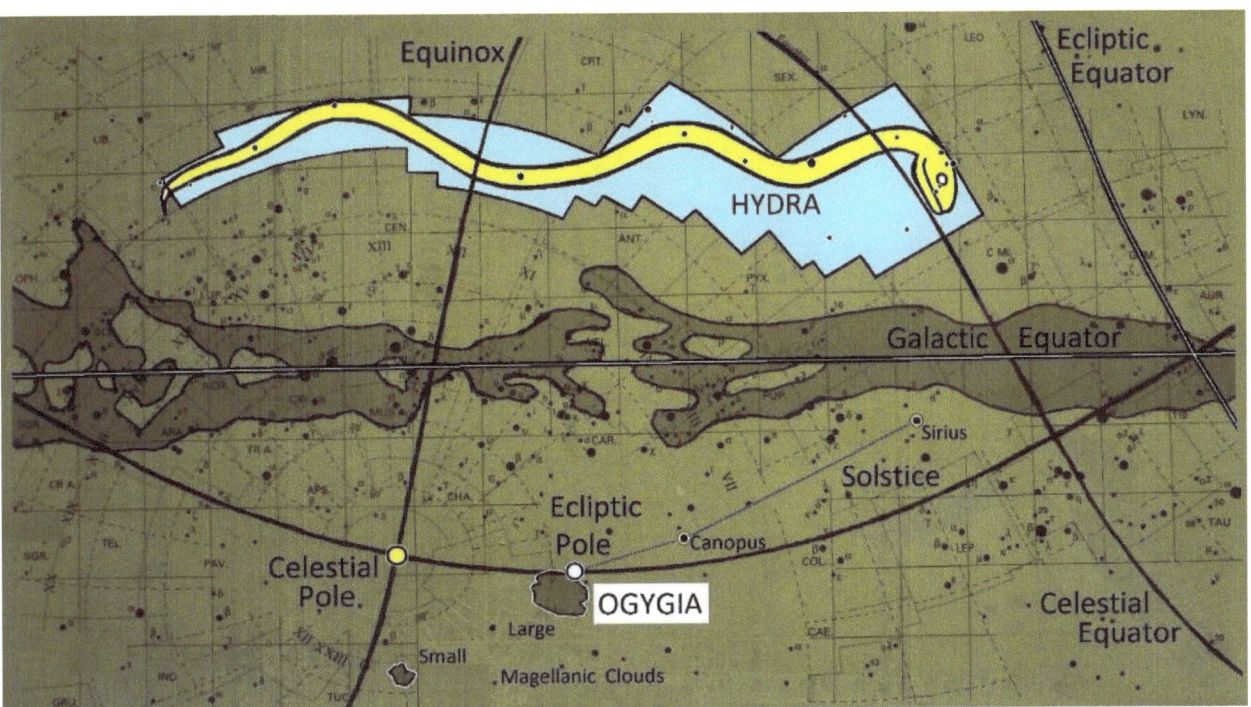

....

Chapter 13: The Three Dragons, 2) Draco

For the other 2 serpents in the heavens. One of them has had ancient mythos told about it that encode the greatest of all mysteries... ever. That is DRACO, the Dragon; curled around the North Ecliptic and Celestial Poles. The more ancient version of Draco had wings. And one of the stars of the wings is actually our current pole star, Polaris... now the tail tip of the Little Bear (TCSotN, Vol. II).

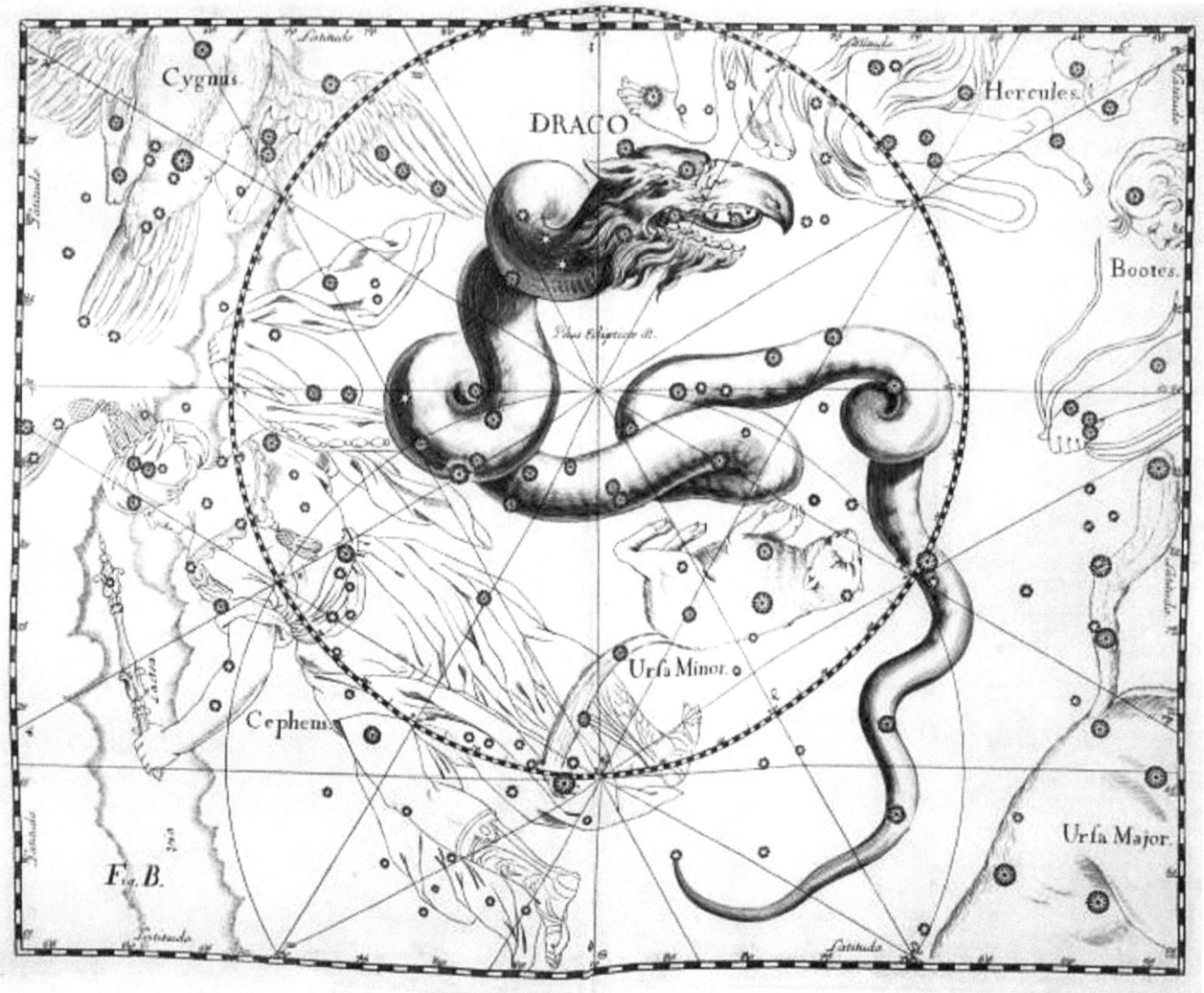

Chapter 14: The Constellations of the 'Beast

It is this DRACO, together with the BEAST, that each has the motif of the seven heads. And as I said, it encodes the greatest of all mysteries. Quite the buildup. But I stand strong behind my statement, as I know the evidence I LATER present lives up to this. Thus; more later, on Draco the 2nd Serpent.

And the final 3rd Serpent is the one in the vision of Saint John the Divine, that relates to all the other constellations in the vision. The 3rd Serpent points directly to the VERY DAY, THE MALE MESSIAH IS BORN. Again, more later.

Before explaining the other 2 Serpents more fully; I will uncover another important clue on the Astronomy meaning of the BEAST. But not the 7 heads part of the vision; so this is 'not' the part of the motif that relates to DRACO as the serpent. This part about the BEAST stands alone.

Just after *Revelations* introduces the Dragon; it introduces the BEAST.

> "John saw the Beast rising up from the sea, having seven heads and ten horns, and upon his horns were ten crowns, and upon his heads was the name of blasphemy. It was like a leopard, with feet like the feet of a bear, and had a mouth like a lion."
> *Revelations 13:1-2*

This clue about the BEAST points to a very specific location in the starry sphere. The body of the Leopard does not refer to a constellation, but instead it is the starry sphere itself. The spots represent stars. But the other two parts of the Beast are very well known constellations. Leo the Lion, and Ursa Major the Big Bear. And in this region of the heavens between the constellations of Ursa and Leo there is a very important stellar feature. The North Galactic Pole.

By pointing this out, the ancients showed that they were aware of the fact that the Galaxy did indeed have an Axis; which the Galactic Equator was perpendicular to. Just like the Celestial and the Ecliptic Axis' both have their Equators. The area in the starry sphere around the South Galactic Pole was not made into a constellation until recent times; when it became part of the Constellation the 'Sculptor' or the 'Sculptor's Workshop'. I believe that the ancients who were aware of the Galactic Axis, and its North and South Galactic Poles, kept it a secret. And the fact that the ancients knew where the Galactic Axis was located is very important. But the reason WHY it is so important is not explained in this book. That is saved for another book.

According to *Maya Cosmogenesis 2012* (by John Major Jenkins) The Mayan's and the Toltec's had very similar cosmologies. And both cultures used the concept of the zenith sky very much. This is the exact top center of the sky, for wherever a person is located within the tropics. In *Maya Cosmos*, (by David Freidel, Linda Shele, and Joy Parker), it gives evidence that Mayans also understood that our Galaxy had an axis. The Mayan's had a specific glyph, the Ek'Way, to represent when the band of the galaxy was resting on the horizon. And when the visible galaxy was on the rim of the horizon, encircling the earth; the North Galactic Pole is in the Zenith of the sky. This also means that the South Galactic Pole is in the Nadir position. In this position the Galactic Axis stands straight up and down; as the largest World/Cosmic Axis. During our current era of Precession, the North Galactic Pole is on the north/south on the Celestial Meridian at midnight and noon on the dates of April 3rd and Oct. 2nd, by my calculations.

And now back to the Old World and the Constellations of the BEAST.

The constellation of Virgo faces the North Galactic Pole. And in that constellation, one of the hands of the Angel is usually held out in front of the Angel's face. Many times a single finger of the hand points out. Sometimes the finger points back to the face; and sometimes it points upwards from the body.

UNDERSTANDING REVELATIONS BY ASTRONOMY

But my favorite constellations have Virgo's finger pointing at the North Galactic Pole. The Galactic Poles are easy enough to plot out in the sky when you know where the Galactic Equator is. And even in Ancient times when they did not know exactly where the Galactic Equator was; it was easy enough to approximate it, and point out the Galactic Poles.

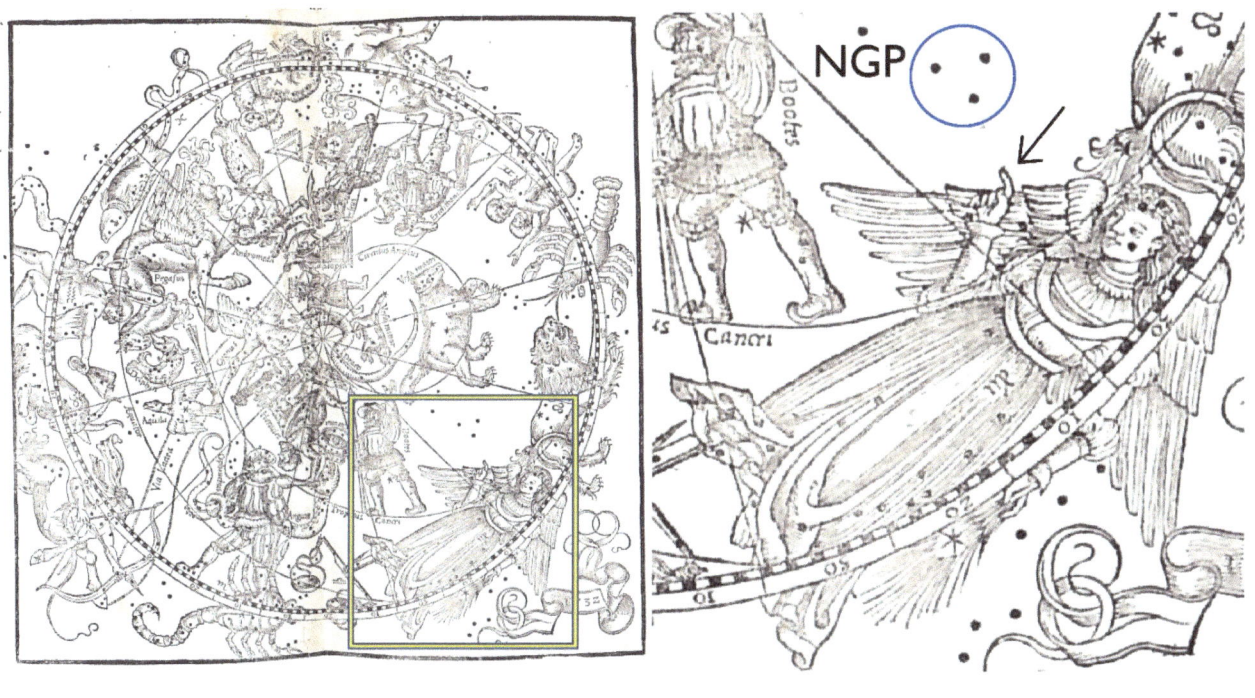

Chapter 15: The Beast & the Woman with a Crown

And it does just so happen, that there is a constellation in this area of the North Galactic Pole. It's called BERNICE'S HAIR. Its portrayed as the back of the head of a woman, with long hair. And often a 'coronet' is on her head. A CROWN. The constellation was officially added in the 3rd century BC. And is dedicated to a consort of Ptolemy of Egypt, who cut her hair off and offered it to the Gods. It also had been referred to as Ariadne's Hair.

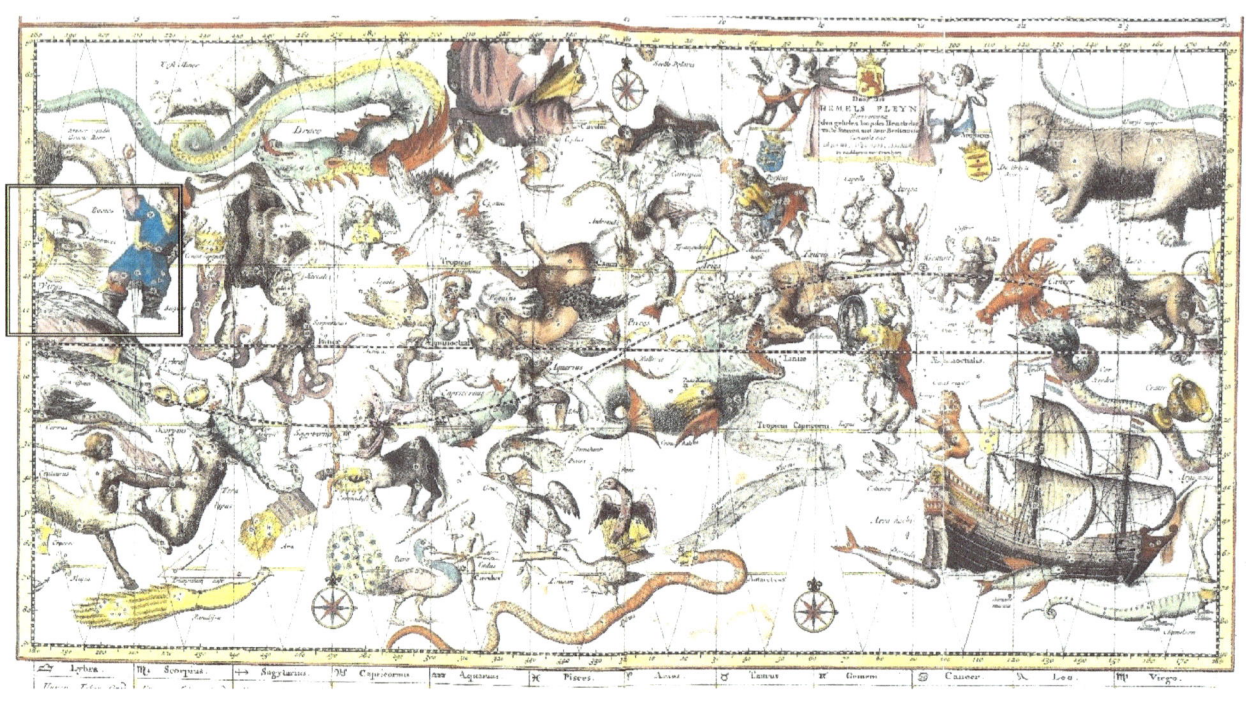

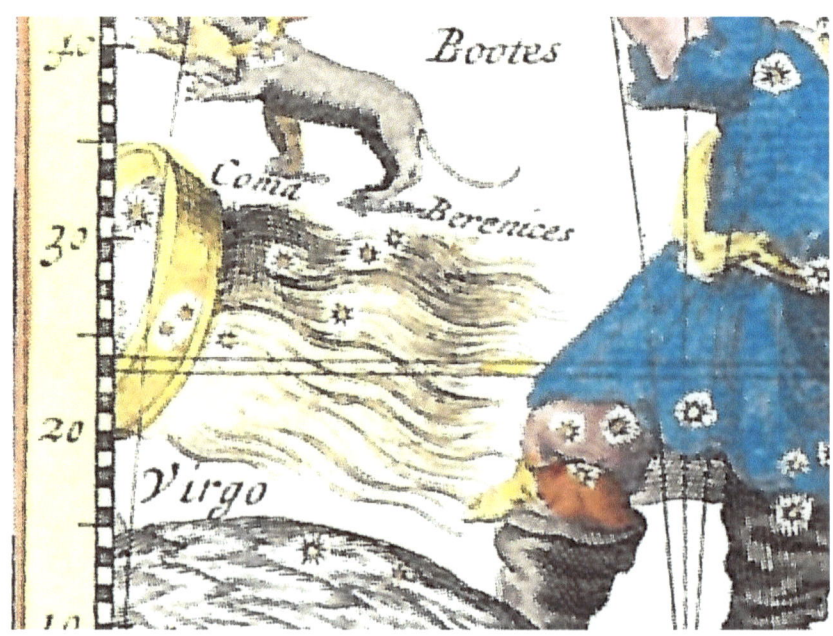

And while there is a Coronet/Crown sometimes shown in the representations of Bernice's Hair. There are only 2 stand-alone Crown constellations in the heavens. The NORTHERN CROWN and the SOUTHERN CROWN. But... the Northern Crown is said to be the crown of Ariadne. Which also ties the Constellation Bernice's Hair to the Northern Crown Constellation.

And by having the Queen's head tied to the BEAST. It is also one aspect of the visual image in *Revelations* from the Woman riding the BEAST. She is on top of him as in both are in the same region of the sky, overlapping.

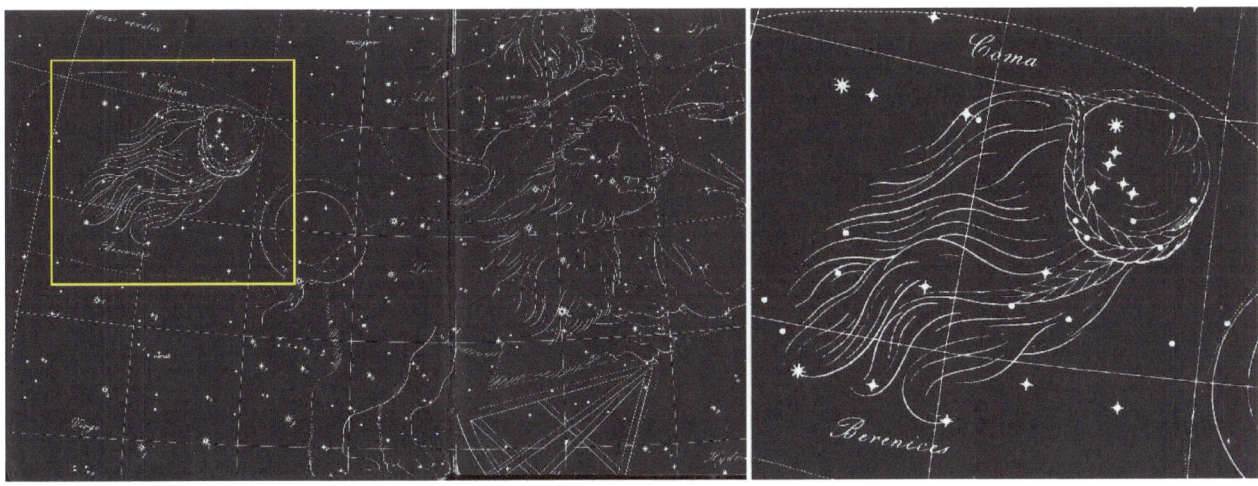

Chapter 16: The Three Dragons, 3) Ophiuchus

Now to the rest of the constellations of the vision. These are the clues, the breadcrumbs on the trail, that Saint John the Divine Revelator left for all.

1) The SERPENT and the SERPENT'S TAIL
2) the CROWN
3) the BABY
4) the ROD OF IRON
5) The WINGS of the EAGLE

And there is a place in the starry sphere where all of these constellations are tied together. The third serpent in the heavens is part of the ancient constellation called OPHIUCHUS. And as I said earlier; this is the Serpent that reveals the very birthday of the Male Messiah. Remember, this is the 13th constellation that touches the Ecliptic Equator. And Ophiuchus is... the SERPENT BEARER. A man holds the serpent in both hands at his side, and the Serpent's body goes either around his waist or between the man's legs; and then through his fists. The head raises up to one side and the tail raises up on the other.

And just above this Serpent's Head is the Northern Crown. This is the only Serpent of the 3 that has a CROWN on its head.

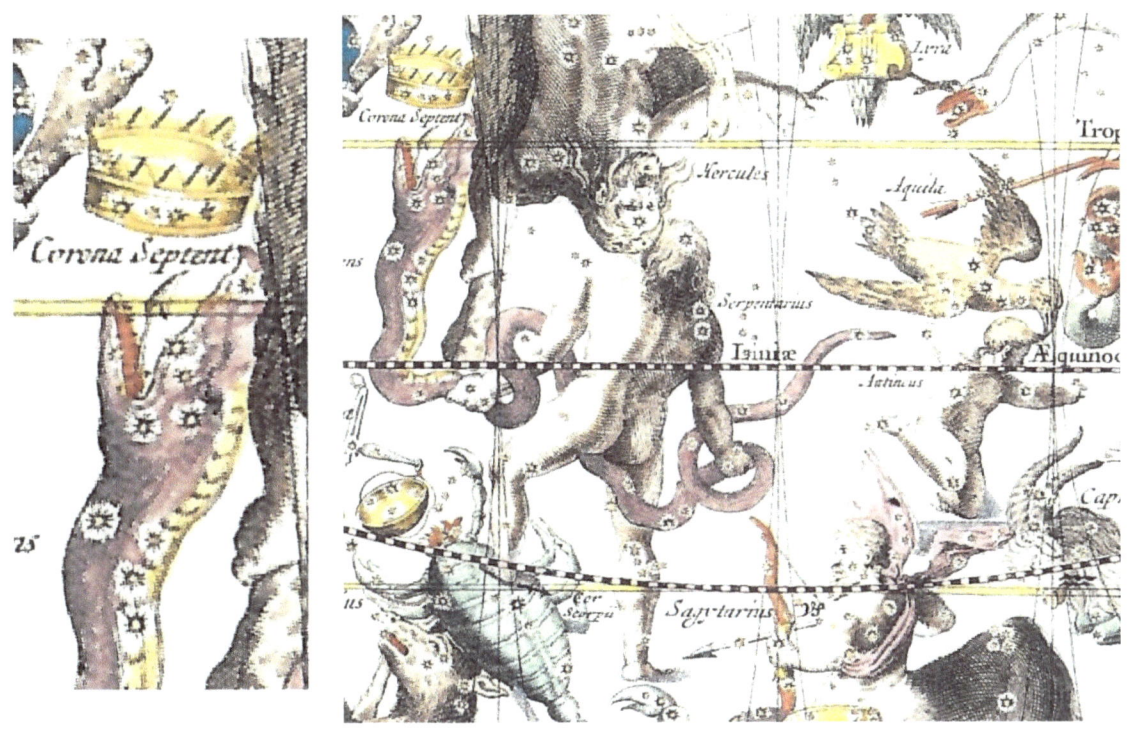

Chapter 17: The Serpent's Tail & the Eagle's Wing

Just at the tip of the Serpent's Tail, is the lower part of the constellation of AQUILA, the EAGLE. Some drawings of the Eagle that I have seen, have the Eagle facing down, or south if you will. This puts the beak closest to the Serpent's Tail. But mostly it is shown sideways, facing away from the Serpent Bearer; with the lower wing being closest to Serpent's Tail.

And that is why it is the ~Serpent's Tail~ that cast one third of the stars from the heaven, and not the Serpent's Head. It is the Serpent's Tail that connects to the Eagle's Wing. And there is something in that region of the Starry Sphere that specifically pinpoints the Birth Day of, "the Man-Child, who shall rule all the nations with a Rod of Iron". So, It is the Serpent's Tail that ties in the day and month with the year of the Male Messiah's birth (from the 1/3 of the stars being swept to the Earth by the Serpent's Tail).

....

Chapter 18: The Male-Child Messiah

And where is the BABY in the heavens, the Male Messiah himself? The only constellation that always has a baby, or toddler male, to adolescent male; is JUST BELOW, where the Eagle's Wing touches the Serpent's Tail. It is sometimes called Antinous, and sometimes called Ganymede. It has been around, -at least- since the time of the Roman Emperor Hadrian in the 2nd century AD. He was completely in love with a young boy and wanted to name it after him. Likewise, Ganymede was the cup-bearer of Zeus in the Greek Mythos. But there are mythos telling that Zeus liked to have sex with the Young Boy. Perhaps it is for these reasons the constellation is not seen in many constellation maps of the heavens.

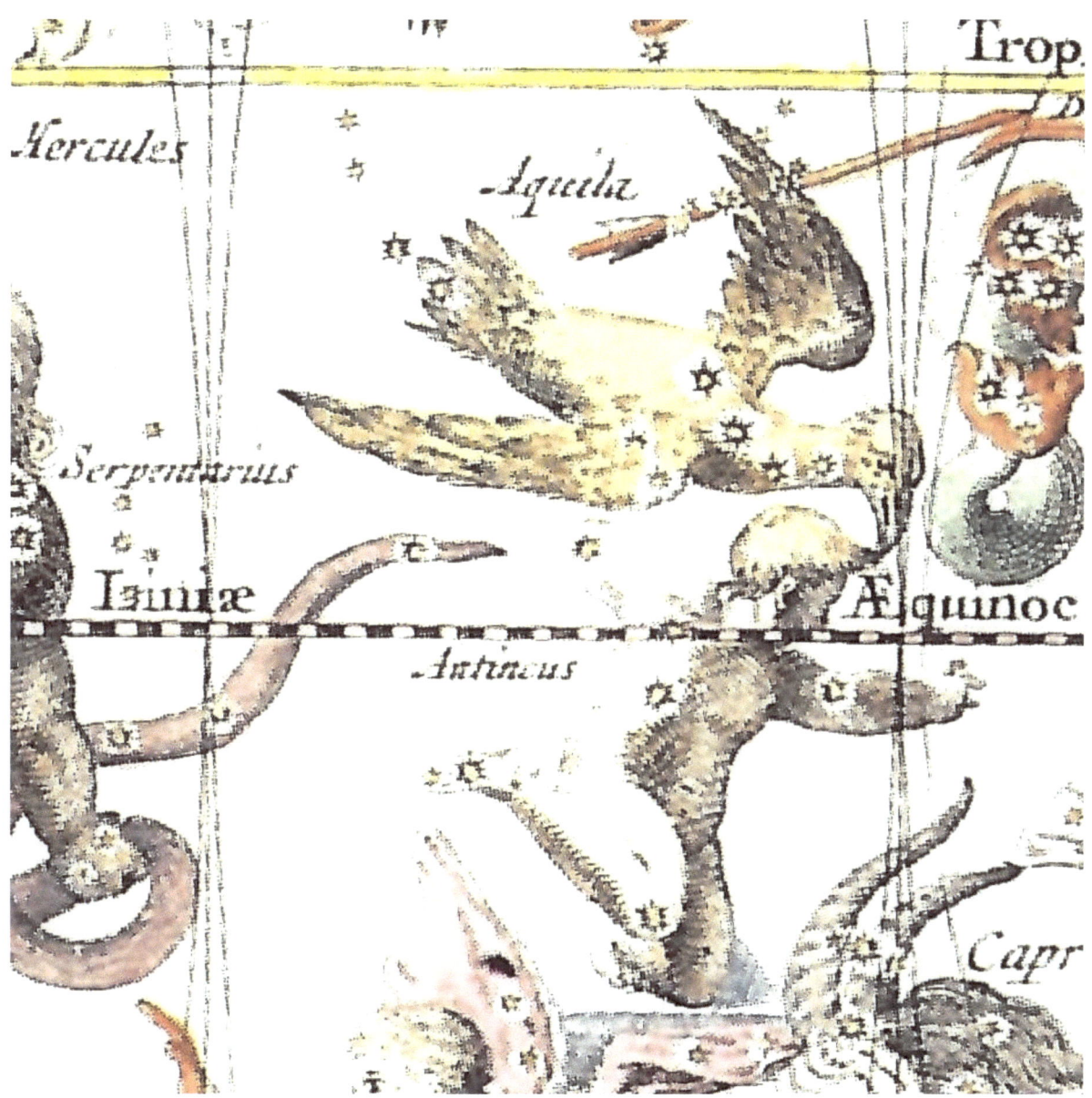

But I think also there is another reason as to why the Male-Child, was often left out of constellation maps. I think the Child/Baby Male was known to be associated with the coming Messiah. And because of that, its meaning was purposefully confused; and the constellation was kept a guarded secret.

In ancient Alexandria, there was the Mystery Religion for Serapsis and Isis, of the Egyptians. Serapsis was like an Osiris, as the husband of Isis. Isis was said to give birth to HORUS on what was the accepted and expected birthday for the dying and resurrected gods. That was December 21, the Solstice. And THEN, on the day that is now January 1st, Isis gave birth to baby AION, the Messiah of the 'Great Year' and the Galactic Alignment. For the early Christians, this was the original 12th night; which was adopted from the Mystery Traditions. It was the 12 days after the December Solstice to include the Solstice itself, January 1st. The Dec. 21 Solstice was USED for the birthday of Jesus Christ; just as it was for countless other Male -dying and resurrected gods. It was expected of him, whatever his real birthday was. Thus, Constantine decreed it so, when he unified Christianity under his rule. But it was already a popular belief.

As many know, Julius Caesar created a calendar. It did 2 things. -1) The emperor MOVED the first day of the year from MARCH EQUINOX; which used to be known as March 1st, to what is NOW January 1st. And of course, this moved our modern New Year from its old calendar date around Jan. 20th-ish; which was re-numbered to JAN. 1st. -2) The older calendar of 365 days was replaced by a new system of 365.25 years around 46 BC; because it was 'discovered' or 'no longer kept secret'- that the entire year was actually a little longer, by a ¼ of a day. Almost certainly, this had been known by the Egyptians; 'before' it was so proclaimed by the Roman Emperor; and made common knowledge to all. The Egyptians used the number 36,525 openly in their Mythos.

And, I believe all the Mystery Religions that held knowledge of Precessional Cosmology knew this. And to adjust for it, the Leap Year was invented; to catch up on the ¼ days. BUT THEY WERE WRONG! It was not quite 365.25. We know now, the tropical year (from equinox to equinox) is bit less than that at 365.2422 days. What that meant was; that through the centuries, the Winter Solstice moved from the calendar date, Dec. 21. It became the 22, then 23, then 24, 25... for a total of 12 days. And during the Papal reign of Gregory XIII in 1582. Finally, the year drifting had been calculated, and the new, more advanced, leap year system was more correctly determined. And, after the correction, the Solstice was moved back to December 21. BUT the birthday of Jesus was moved to on the 25th, rather than moved back to the 21st. This was done, TO SEPARATE JESUS from all the dying and resurrected gods.

....

Chapter 19: The Shield of Kings

In 1690 an astronomer named Hevellius created several constellations in the starry sphere. He made a SHIELD constellation just below where the Eagle's Wing and the Serpent's Tail meet. On that Shield he put a large cross emblem. And as we shall soon see, there was a very good reason for having that cross on the Shield, because of its position in the starry sphere.

As you look at the star charts I present here, there are 2 different ways all the various constellations are portrayed. They were either drawn, as we see the stars from the Earth, looking towards outer space. Or they were drawn mirrored from that, as if we are in outer space looking at the Earth with the stars between us and the planet. Some constellations even got redrawn in other ways through the centuries, some face different directions. You can see how Antinous, the Male-Child constellation, now holds a bow and arrow.

When Hevellius made his constellation charts he definitely put both the Eagle and the Male-Child on the same plane, metaphorically tying them together. The arrow in his bow reinforces this. The Shield aligns vertically like Sagitatius below it. The Shield Constellation was internationally approved in 1922, with the current 88 constellations. But, they got rid of the Male-Child, hiding him again.

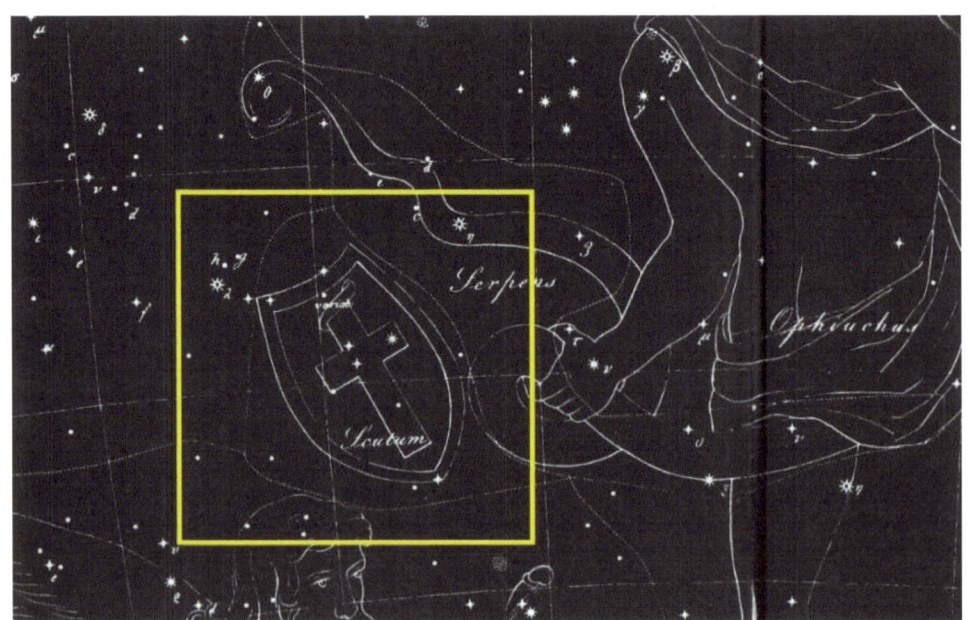

Chapter 20: Who is the Serpent Bearer?

And finally, what about the WOMAN who gives birth to the Man-Child? Who is it? Virgo is a good bet, true; and we have explored the Angel's connection with the constellation Coma Bernice, that marks the North Galactic Pole. But there is another constellation that is a better candidate. The Serpent's Tail is part of the constellation Ophiuchus, named the SERPENT BEARER. It's shown with a man holding the Serpent.

If Ophiuchus was a Woman, it would just tie up all the clues very neatly. After all, *Revelations* says, "the Dragon stood in front of, the Woman about to give birth". This is the only Serpent of the 3, that clearly interacts with a human.

I believe, in very ancient times, this constellation was definitely a WOMAN. Because the Wise Mistress of the Underworld and Death, was ALWAYS associated with the Serpent. A middle ages star chart called, "Aratus' Map of the Heavens"; by, that Romance Era Artist; does indeed portray the constellation as a female.

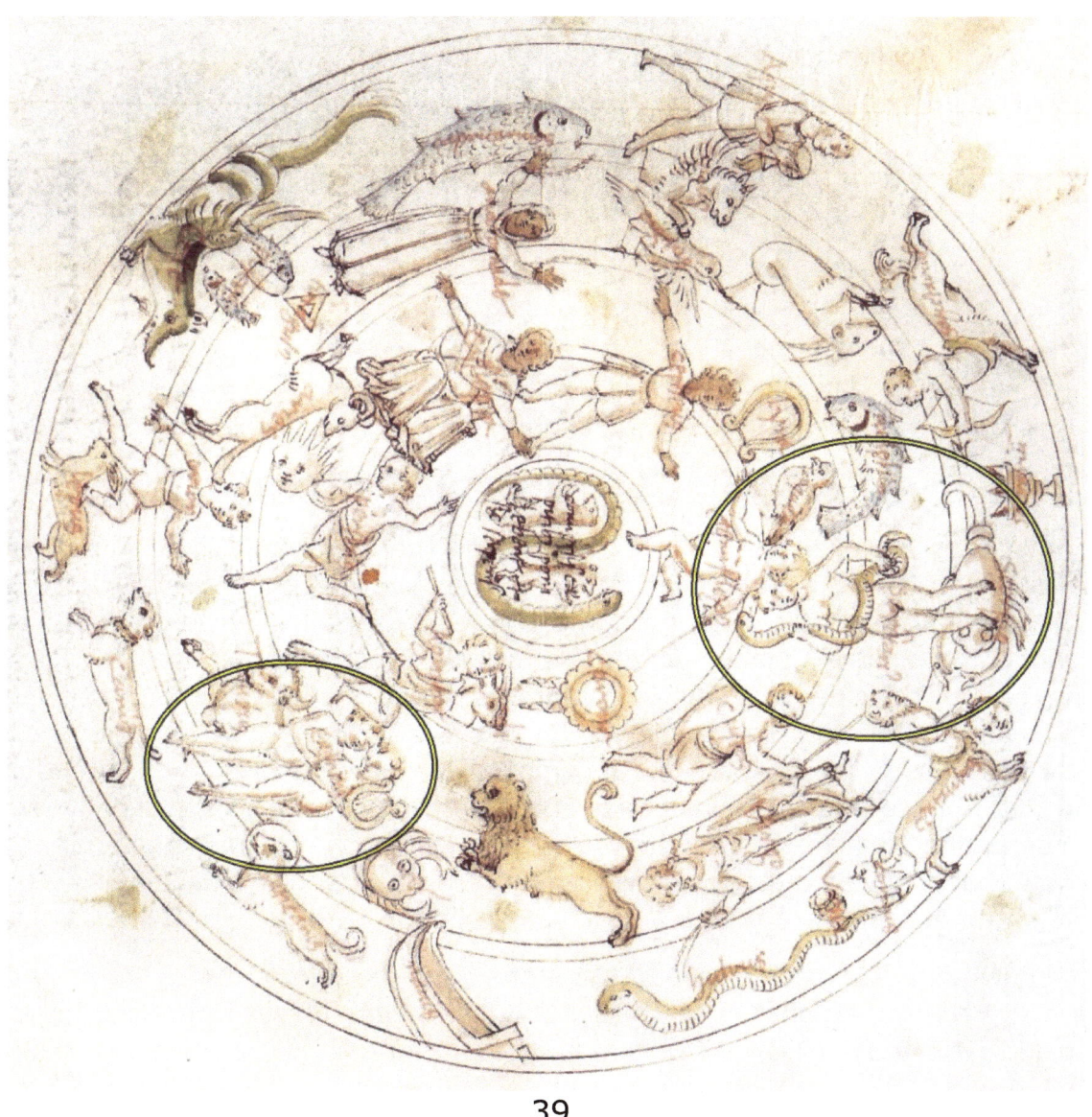

UNDERSTANDING REVELATIONS BY ASTRONOMY

The Romance Era was filled with people, who had learned the secret Templar knowledge. All of the constellation figures of this star map are young o adolescent and most are clothed. But two constellations, Gemini and Ophiuchus; show the genital areas. And while the Gemini twins are shown with; what appears to be little bumps, suggesting male genitals; Ophiuchus is shown very smooth, like an adolescent girl.

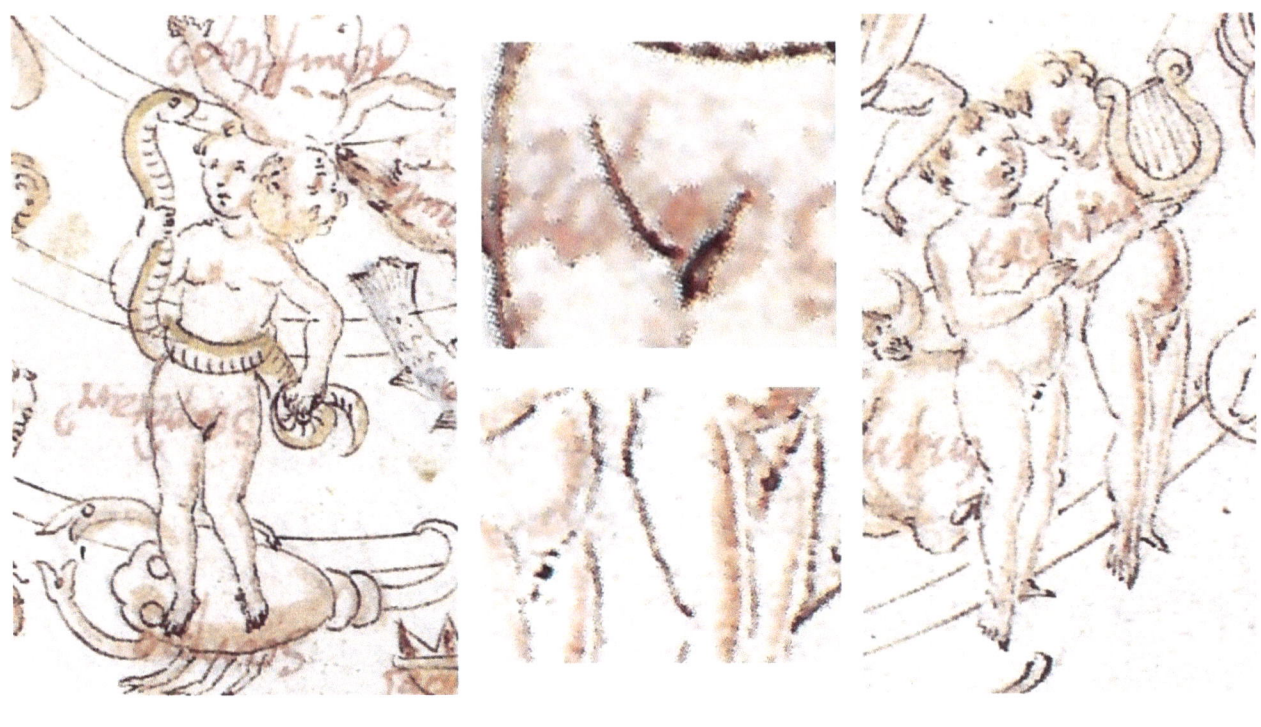

Chapter 21: The Rod of Iron

We have explored the vision of Saint John the Divine in *Revelation 12*; and I have identified the constellations that were seen by John the Revelator. We know what year the Male Messiah was born in, 1966; and that all 3 Serpent Constellations are connected to the Galactic Alignment of the Precession of the Solstices and Equinoxes. But the clues of the vision point to one specific area of the starry sphere; which is where Serpent's Tail meets the Eagle's Wing. The WOMAN of the vision has several possibilities, Virgo; or a female Ophiuchus. And the CROWN is directly next to the Head of the Snake of Ophiuchus. The Crown was used to specifically identity which serpent of the three is for the male messiah's birth day in the solar year. And we know that the area directly below the Serpent's Tail and the Eagle's wing is also important as this is where the Man-Child constellation is.

So why is this specific area of the starry sphere so important? We know from St. John's Vision, that what is so special about the Male-Messiah, is what gives him his power... It is the ROD OF IRON, that he holds; and with it "He rules all the nations of the earth." Yet the Rod of Iron is not a constellation, but it is a stellar feature. The Rod of Iron is mentioned 3 times in St. John's *Revelations*, and also in the Old Testament. It lies along the Milky Way. But, the Rod of Iron is not 'all' of the Milky Way. It's just a specific part of the Milky Way.

The Milky Way was 'seen' as a different thing in different cultures. It was seen as a tree, a road, a river, a canoe. And also, it was often seen as a long-snouted beast'; like a Crocodile in Egypt; or a Wolf; or an ASS, like in the Middle East. And the two spots after the mouth are the eyes.

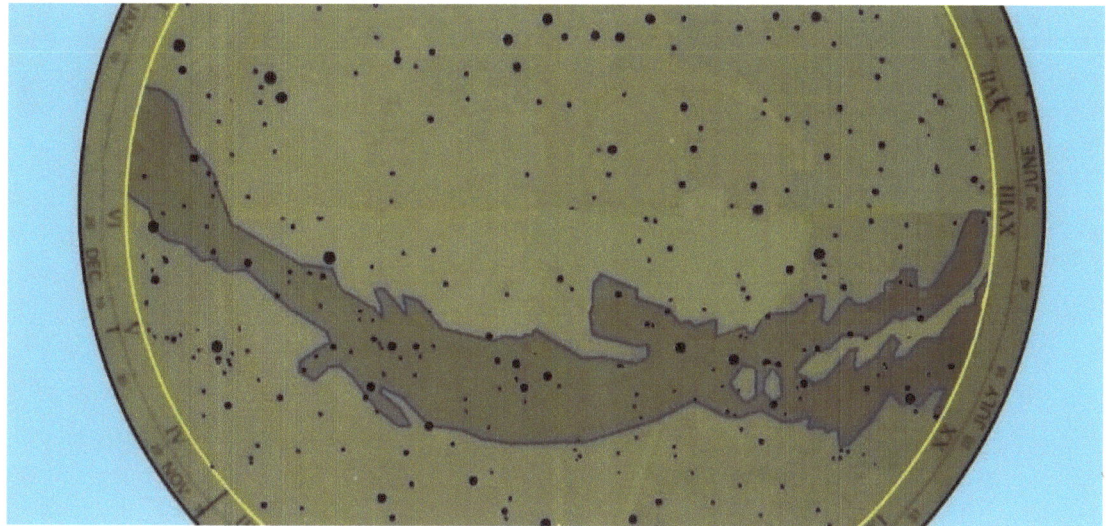

There is a very thin line in the middle of that thick band of the Milky Way stretching around the starry sphere... the Galactic Equator. And that's where the Rod of Iron is. The Galaxy has an Equator that spins out perpendicular to the Galactic Axis, with its North and South Galactic Poles.

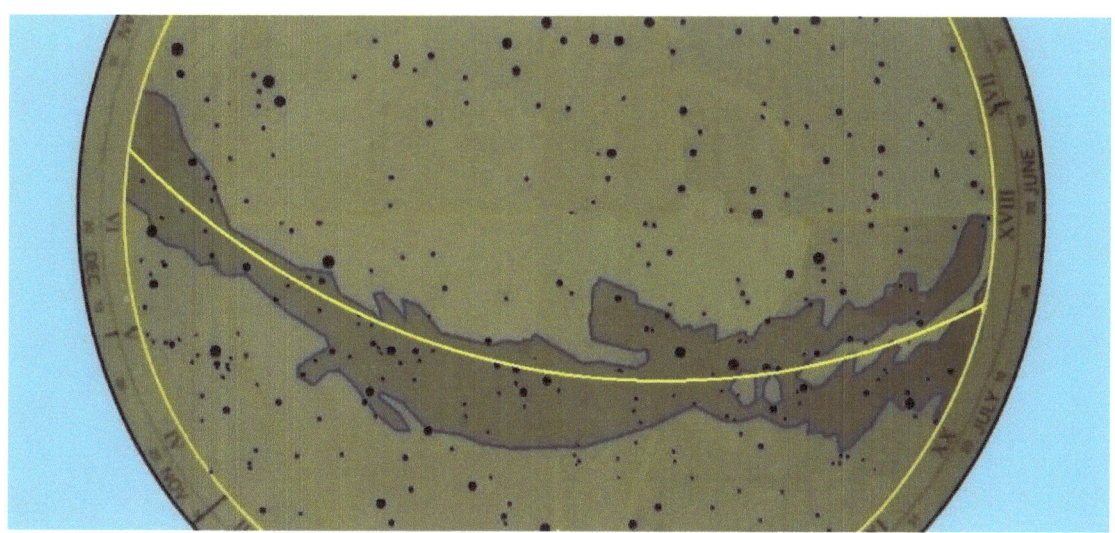

The area that is the Rod of Iron, is just a small portion of the Galactic Equator. Its top starts where the Celestial Equator crosses the Galactic Equator; and of course, that's where the Eagle's Wing touches the Serpent's Tail. The Rod runs down the Galactic Equator, crossing the 23 and 1/3 degrees to reach the historic Galactic Center. So, the Rod of Iron goes down and passes beside or through the Man-Child constellation. And also, the Rod of Iron goes through the Shield Constellation with its Cross emblem; and it continues down a little bit more, ending at the Ecliptic Equator. This area is also… the JAWBONE of the ASS, from Samson, in the *Old-Testament* book of *Judges*.

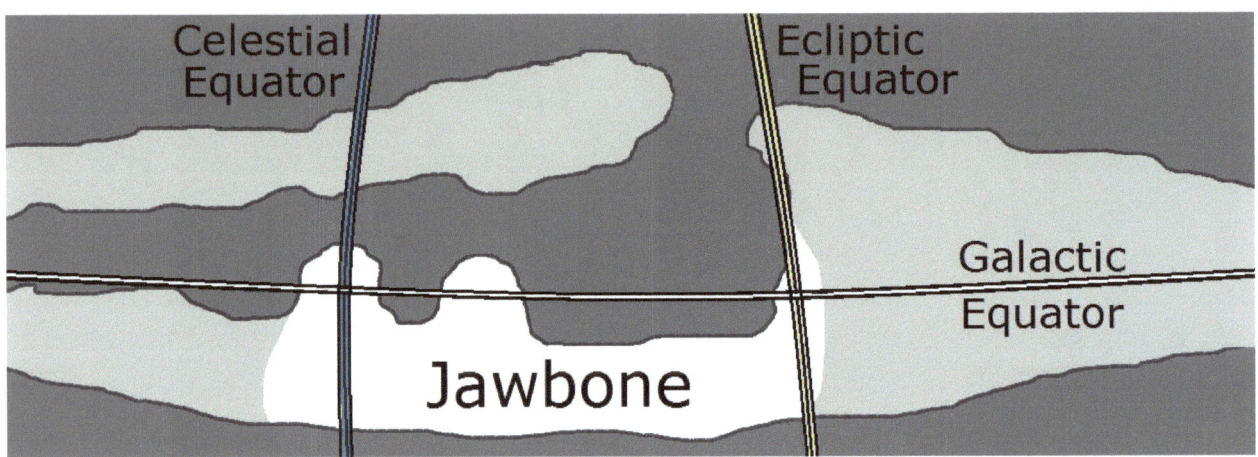

Even as the Galactic and Ecliptic Equators are relatively un-moving; the Celestial Equator wobbles like a spun coin just before it rests on the table. But as the Celestial Equator wobbles around the Ecliptic Equator, it's placed halfway through, the top and bottom of the tilted Celestial Equator. It's the top and bottom of the Celestial Equator that creates the Tropics of Capricorn and Cancer. What's important here is that the position of the Celestial Equator is time based. It's constantly changing place, during its one revolution wobble of the 'Great Year'.

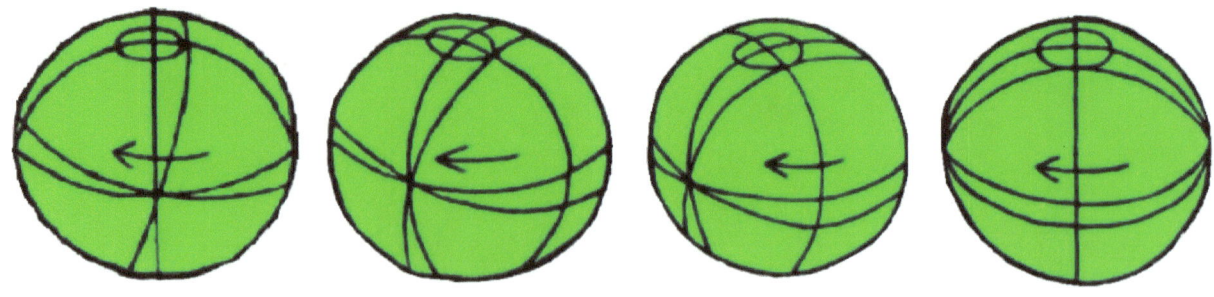

And the Rod of Iron is of course here and now in 'our' time. It is about the Galactic Alignment, the end of times as we know them; and it involves all three equators. Remember that at four equal spaced times, the 2 Solstices and the 2 Equinoxes cross the Galactic Center for a 72-year Golden Era; and there are 6408 years between each of the 4 passes. 72 + 6408 = 6480, 6480 X 4 = 25,920.

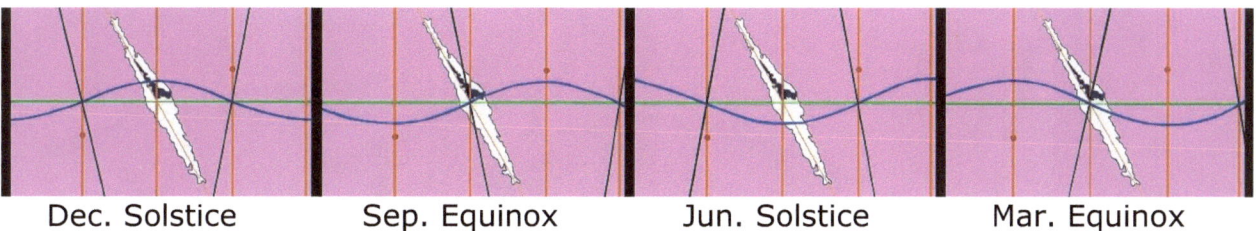

Dec. Solstice   Sep. Equinox   Jun. Solstice   Mar. Equinox

What makes this the Galactic Alignment, is that the December Solstice is now aligned with the Galactic Center. That's the 1 out of those 4 that stops and starts the whole 'Great Year'. And since it's the December Solstice, this means that the Celestial Equator is the farthest it ever gets NORTH, from the crossing of the Galactic Equator and the Ecliptic Equator (the Galactic Center).

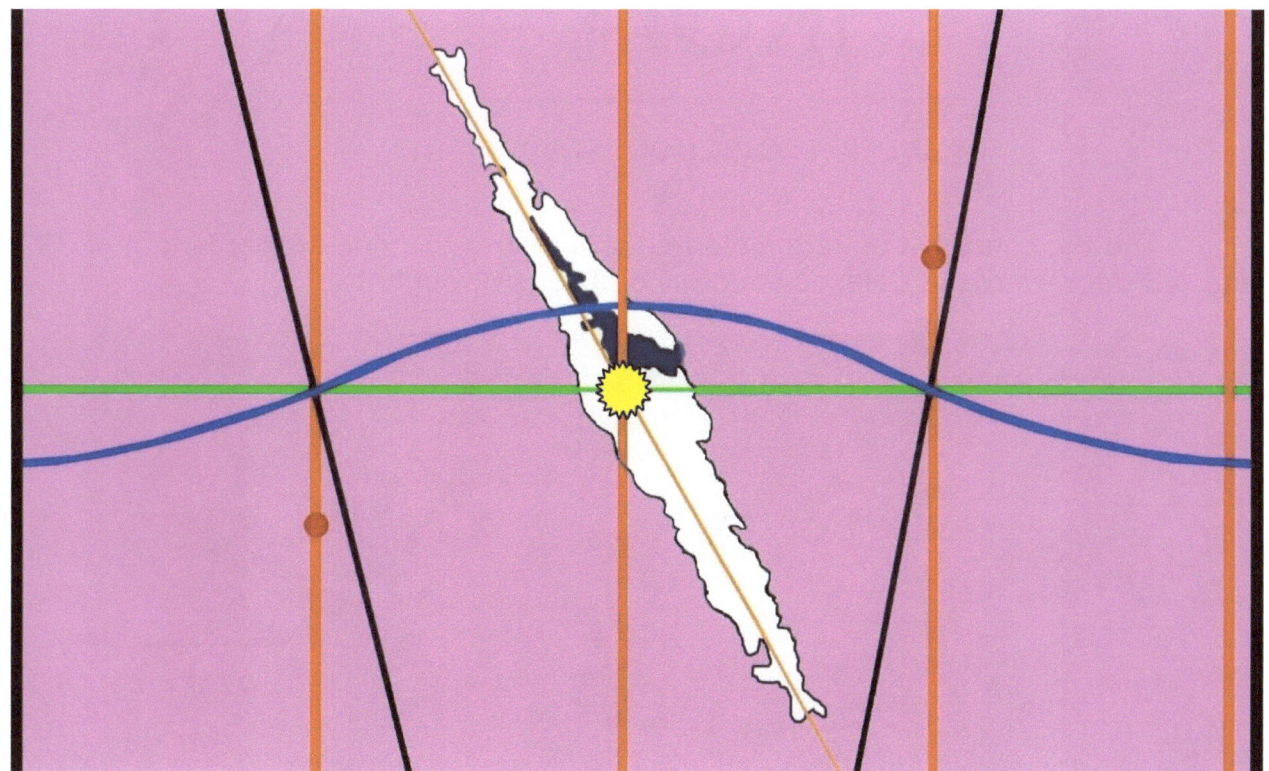

Right now, is the Galactic Alignment, the End of Times as we know them. So, the Earth's Celestial Equator has moved to be 23 and 1/3 degrees north of the Galactic Center; where the Eagle's Wing touches the Serpent's Tail. And the space below it, running south along the Galactic Equator, is the Rod of Iron.

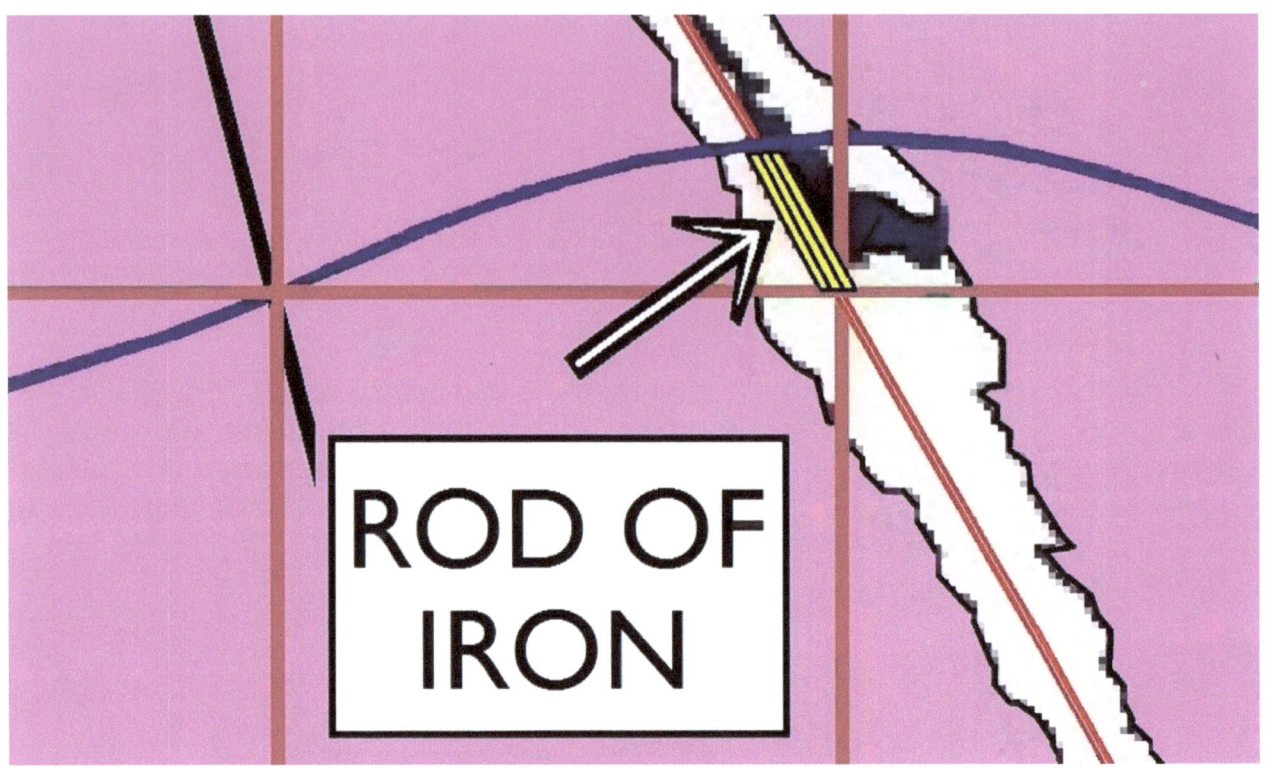

CYLINDER PROJECTION STAR CHART

And the constellation of the Man-child is just to the side of that length of 23 and 1/3 degrees of the Galactic Equator, or it runs through it.

And right on that area of the Galactic Equator is the Constellation of the Shield with its cross emblem. So of course, the Cross emblem represents the cross made by the Celestial Equator and the Galactic Equator; However, the Cross of the Shield is a little south of the true Cross made by the Celestial Equator crossing the Galactic Equator.

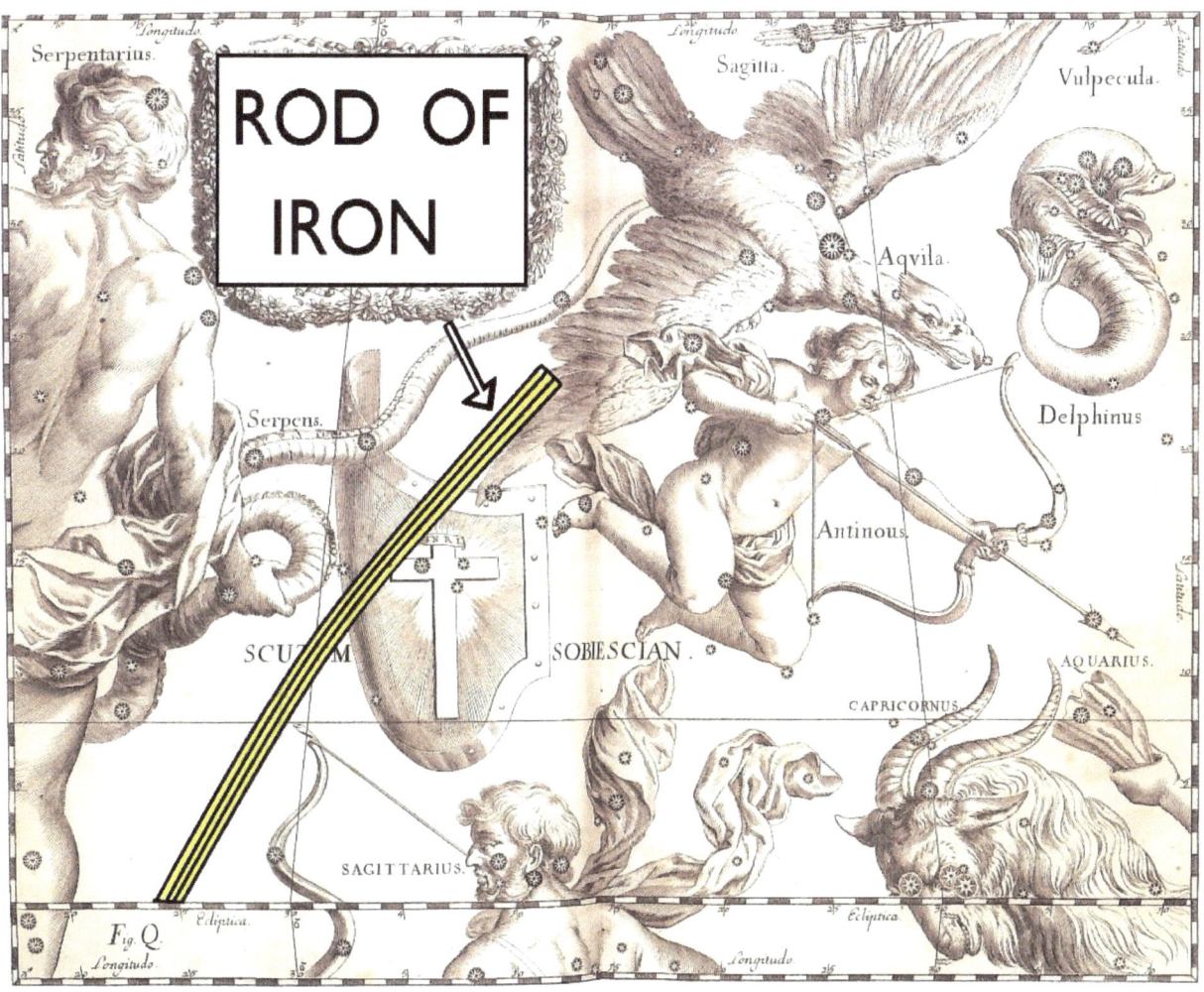

Remember, as I said earlier; the constellation charts that I display have two orientations of pictures; as they were made throughout the various centuries. And these 2 sets of images mirror each other. Either we look from Earth, or we look from outer space. As you can see from the various star charts for the Rod of Iron that I have presented here.

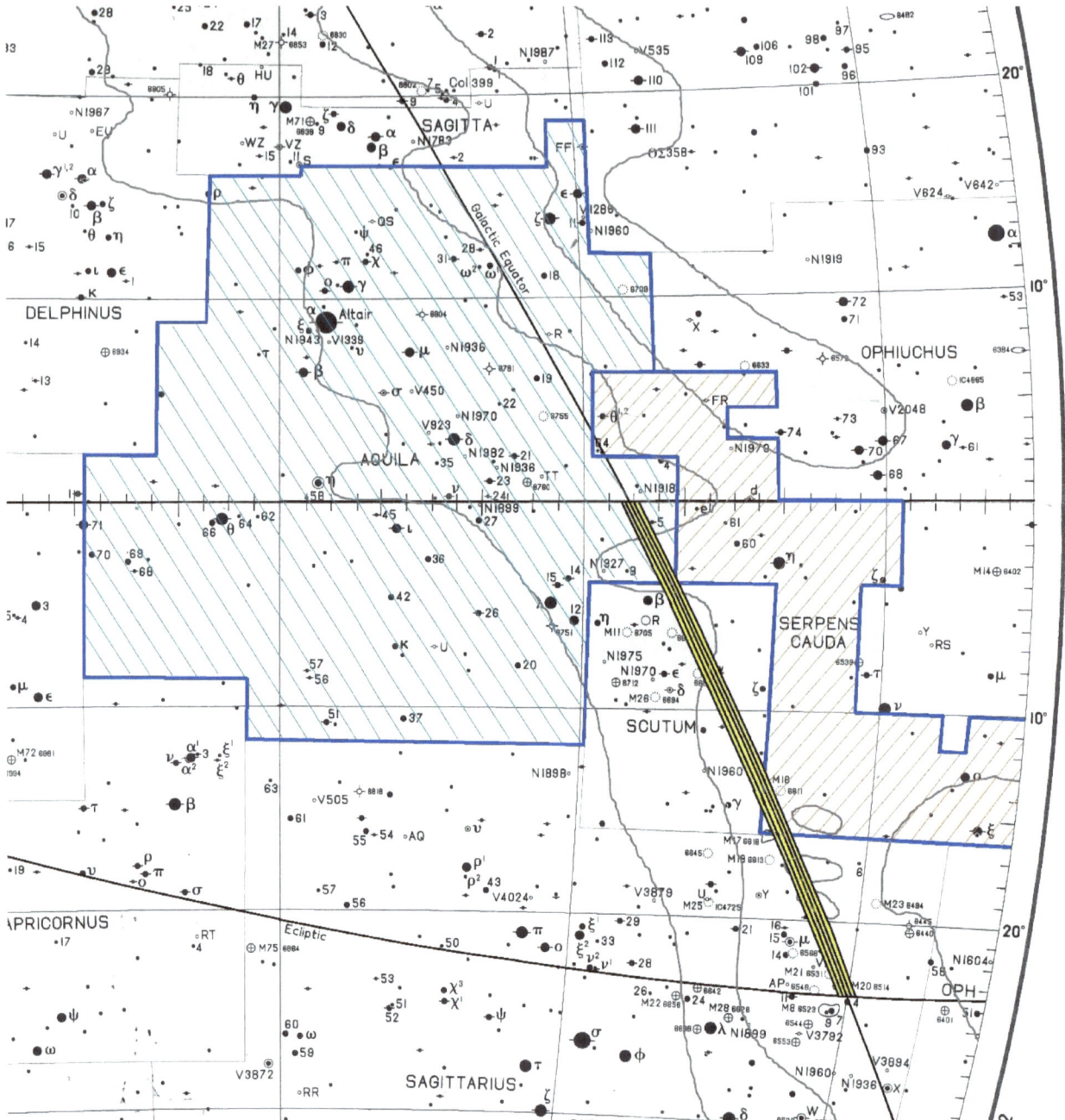

So now we know that this crossing of the Celestial Equator and the Galactic Equator holds the birthday of the Male Messiah. It is the December Solstice alignment, the 21st; when the Solstice 'SUN' is in conjunction with the Galactic Center. And January 1st is when the Celestial Equator crosses the Galactic Equator, at the north/south Celestial meridian, during the noon sun.

But, the Celestial meridian, also has a midnight conjunction, 6 months after the noon-sun conjunction. The midnight Celestial Meridian nexus of the Ecliptic

Equator and the Galactic Equator, at the Galactic Center; is on the Jun.21st Solstice. And the Celestial Equator crossing with the Galactic Equator on the midnight Celestial Meridian is at July 3rd; which is just 6 months passed New year. As Jan. 1st to July 2nd is 183 days; July 3rd to Dec. 31st is 182 days, plus a 1/4 day.

And technically each of those 2 days, both July 3rd and Jan. 1st; is a candidate for the day and month of the Male Messiah's birth day in 1966. I believe it is almost certain that, this is why Julius Caesar moved the new year from Spring Equinox to January 1st. Just like the cult of Serapsis from Alexandria taught that Horus, the god of the year, is born on December Solstice; and that Aion, the god of the Great Year, is born on January 1st. As well, Julius Caesar named July after himself. And this is why, I believe, the US Declaration of Independence was first signed on July 4, even though it was finished on July 2. To pinpoint July 3rd between them. And there is more proof later that the Masonic Founding fathers knew about Precessional Cosmology.

....

Chapter 22: The Great Mystery of 108

Over and over in *Revelations* it mentions the 7 and 7; as well as the 1/3. And before we decode the all-important 7 and 7, we shall explore the motif of the '1/3'.

In humanity's vast collection of cultures, the number 108 is prominent. The fact that the number 108 is tied to the Precession of the Equinoxes and Solstices, is very well known. The reason why this is so, to my knowledge, has never been publicly explained; perhaps until now.

The mystery number, 108 and its multiples, has been used as a common numerical motif in the various traditions of Archeo-Astronomy. And they are found in many of the world's ancient mythos. The other 'famous' number in our field is 72. But the meaning and importance of 72 is very well known; it's the number of years it takes to move just one degree in the heavens, during the Precession of the Equinoxes and Solstices. There are 360 degrees, so 360 X 72 = 25,920 years of the Great Year. The 12 Zodiac Sign World Ages each lasts 2160 years; which is 20 X 108. And that's where it starts.

The use of 108 and its multiples has been mentioned in books, to include *Inner Reaches of Outer Space*, (by Joseph Campbell; p.7-10). As well as by the magnificent text *Hamlet´s Mill* (p.7 HM). And by authors like Graham Hancock in *Fingerprints of the Gods*. These sources make note of 108's use in the mythos of the vast expanses of time known as 'World Ages'. And the 'World Ages', with the multiples and factors of 108 is connected to the Great Year.

The 3 most often cited examples of this are; -1) the ancient Vedic cycle of World Ages, called the Yugas, -2) the proportions of the Great Pyramid at Giza which is said to be 43,200 to 1, to the whole of the Earth itself (p.48; TMotS), and -3) the Viking mythos named *Grimnismal* contained in the Old Icelandic text, *The Poetic Edda,* (by Snorri Sturluson, trans. By Henry Adams Bellows; p.93); which begins the saga by giving the two brothers ages as 10 and 8 (p. 85 PE); as well as, the often-cited lines of verse with the numbers 540 and 432.

And any search will soon bear fruit from many sources, as to how often the multiples of 108 are found. They are in 'The Kings List' of Mesopotamia by Berossus. There are 108 Gopis, as the wives of Krisna, as well as there are 108 prayer beads for the modern Hare Krisna, and the Buddhist prayer beads, and for the Sikh Mala of the Hindis; and 54 on the Catholic rosary. In *The Egyptian Book of the Dead*, (by E.A. Wallis Budge) The Egyptian path through the Underworld has 12 houses, with each house having 9 guardian deities, 12 x 9 = 108. In the famous site of Angkor Wat, where the book *Heaven's Mirror* (Graham Hancock, Santha Faiia) shows the layout of buildings loosely marks the positions of the stars from the constellation Draco; there are several examples. In the book *The Message of the Sphynx* (Graham Hancock and Robert Bauval) and, again, in *The Fingerprints of the Gods* (Graham Hancock) it covers the 108 numerology. In more modern times the number 108 is still given importance. There are 108 stiches in a baseball. And most telling, there are 108 stiches in a Masonic Apron of the Free Masons.

108 is even being used in Hollywood, like from the high concept television series named *Lost*. And also, by the same writers, the fairy tale based series *Once*. It has been shown in movies like the Sci-fi thriller by Luc Besson, *The 5th element*, and less over the top by the masonic inspired *National Treasure*. The modern practitioners of the Art of the Fugue, encode the symbols of hidden knowledge; into the stories of poetry, books, legends, mythos, songs and now ~movies. And just as their ancient predecessors did, before this modern age; they use this number. And even if they do not know the hidden significance of 108; they do know it is important, so they use it.

As it was noted to begin with, the use of 108 is very well cited. The clue as to its meaning in 'Precessional Cosmology' lies in how the numerology system is used in the 'Great Year' along with its sibling, the number 72. Some modern sources may claim an exact number of years for the 'Great Year'. And, indeed, the ancient Mayans counted out the days exactly as 26,000 years of 360 days each. The Old World of Asia, Africa and Europe specifically put the number of solar years in the Great Year as 25,920. Because that's how long it takes Precession to move through the 360 degrees; at 72 years every degree. Jane Sellers, *Death of the Gods in Ancient Egypt* (p.205) asserts it is 1 degree every 71.6 years. So, it is possible in the field of Precessional Cosmology, that this number was chosen, not for its total accuracy; but for it being connected to 108. 36 X 3 = 108 = 36 + 72

Perhaps if there is a tie between the Old World and New World astronomy systems it will be found in the Egyptian and Mayan use of a special year holding only 360 days. We are told in school that it was the ancient people of Sumer that created the circle of 360 sections, called degrees. We are also told that it was ancient Sumer that gave us the count of time with 60 seconds in a minute and 60 minutes in an hour. And this system of 60 minutes and 60 seconds of a degree was adopted to further divide the space in between each single degree in the circle of 360 degrees.

So there we have our clues as to why 108 is important and just how it is used in Precessional Cosmology. First, there is the numerology system based on 25,920 solar years where every 72 years 1 degree is moved in 360 degrees. Second, the 108 has been used to hold the cosmic serpent. And as a string in a circle, as prayer beads, as the number stiches that surround a baseball, and as a border to encompass the Masonic Apron. Truly it's all about the math. 360 and 108 are tied in math, thusly. 360 X 3 = 1080 = 10 X 108.

Do you see how it works? In 360 degrees we divide 10 degrees into, well... 10 sections. But in a 108 division circle 10 degrees is divided into 3 sections. So, each division of 1 section is now 3 and 1/3 degrees, out of the 360 degree circle. With 108 divisions in a circle, since every 10 degrees (of 360) gets divided into thirds, then a division of 108 for the 360 degree circle goes.... 0 degrees, 3 and 1/3 degrees, 6 and 2/3 degrees, 10, 13 and 1/3 degrees, 16 and 2/3, 20 degrees, and then we have it 23 and 1/3 degrees.

The 108 division more accurately measures the obliquity of approximately 23 and 1/3 degrees. Because in 10 degree division there is only 23 then 24; and there is no way to evenly divide 10 degrees of the 360 degree circle into thirds of 10 degrees. Every quarter of the circle holds 27 divisions of 108 and 90 divisions of 360. Even though obliquity is currently at around 23.44 degrees, this difference is close enough to the 23 and 1/3 so as, I believe, not to have mattered to the ancients.

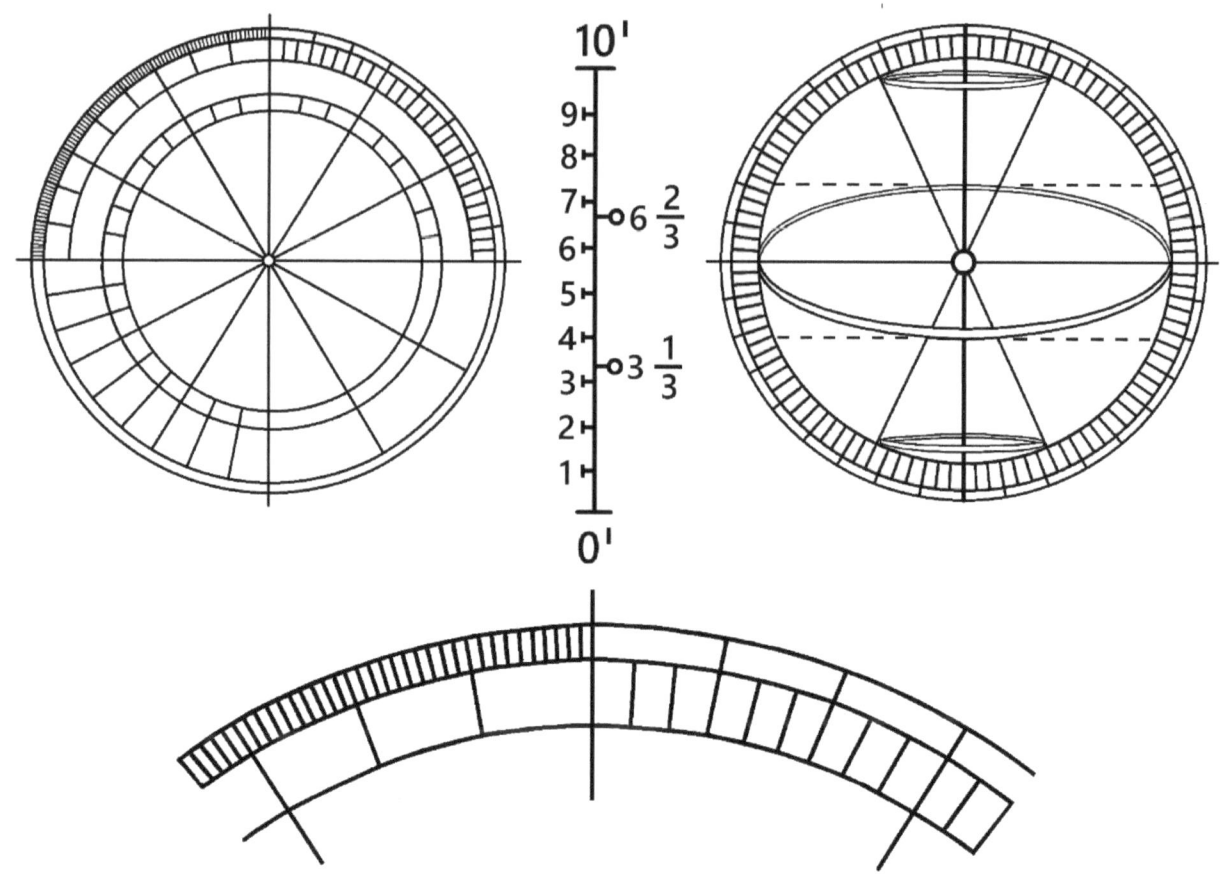

And in the *Revelations* of Saint John the Divine, there is a repeating motif of the 7 and 7 along with the 1/3. When using the 108 sections of a circle, each single one of these sections, is 3.33 degrees of the 360-degree circle. And as you can see from the diagram, 23 and 1/3-degrees is 7 ticks of the 108-degree circle. So, as the Celestial Axis swivels around the centered Ecliptic Axis, on the Circles of Precession; that makes a 7 and 7. The same for the Celestial Equator being tilted 23 and 1/3 degrees above and below the Ecliptic Equator, making the tropics of Cancer and Capricorn. It's the 7 and 7 tied to the measure of 1/3. As well the belt of the Zodiac is said to be 7 degrees above and 7 degrees below the Ecliptic Equator.

PART II - Sacred Geometry & Precession
Chapter 23: Seventh Heaven

The book, *The Celestial Ship of the North*, (by E. Valentia Straiton), is a catalogue composition of stellar mythos; with an entire section devoted to the seven-fold heavenly division. In it, there is ancient reference, after ancient reference, over and over again; about how the both the North Celestial Pole and the North Ecliptic Pole have a 'seven' motif in the world-wide mythos.

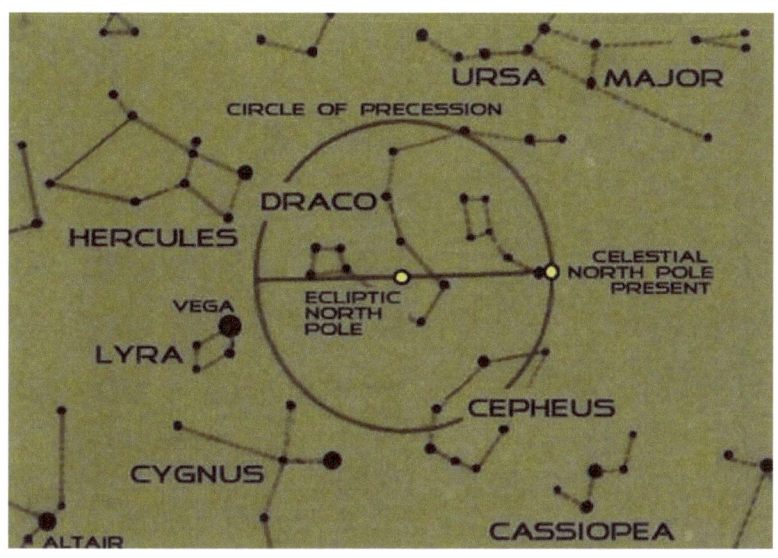

But, as it shall be proven, the North Ecliptic Pole is a red herring. It's the Celestial Pole in the north; in our era, at the "End Of Times - As We Know Them", that holds the 7th heaven.

Many times that true 7-fold point in the heavens, the North Celestial Pole; is tied to 7 concentric circles, one inside of the other, and so on. But even though this is the most common, it's also a red herring. It's the less common division pattern that is the most important. These are the 7 lightning bolts, or the 7 rays of light, that radiate as spokes from their central hub, the North Celestial Pole.

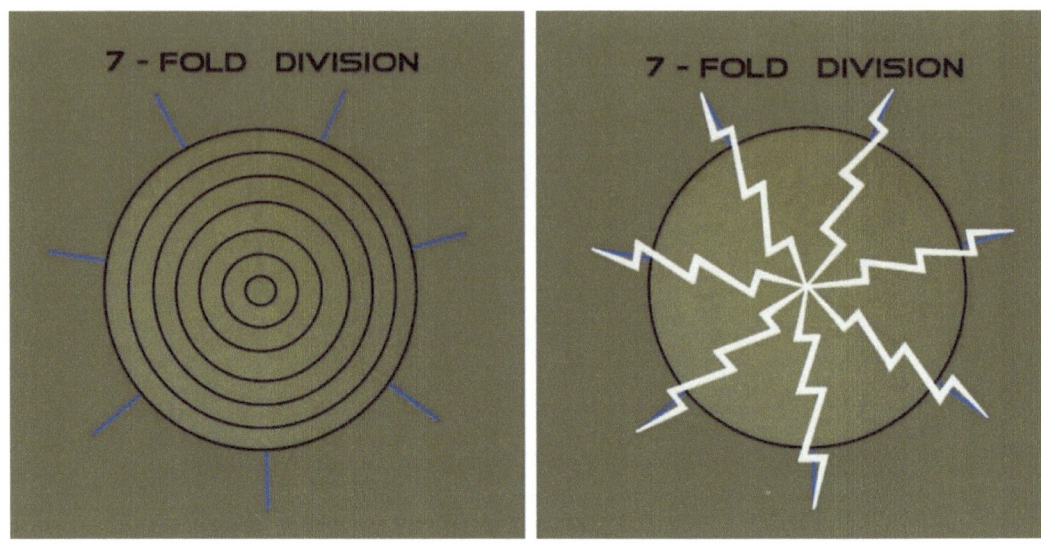

This 7-fold division of the circle creates the 1/7th angle, a 51.428... degree measure.

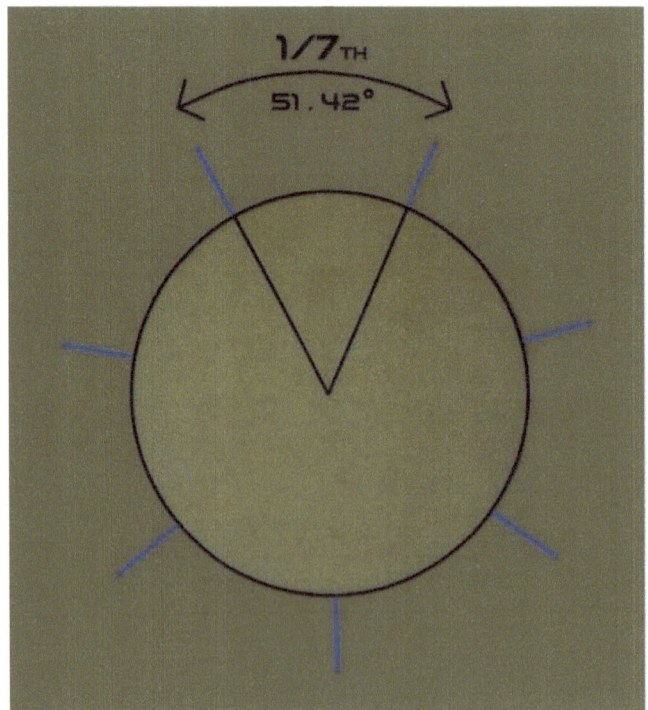

UNDERSTANDING REVELATIONS BY ASTRONOMY

Chapter 24: The 14-fold Division

But even more important than the 7 spoke circle, is the 14-fold division as rays from a center. There are three relevant ways that these 14 pie sections are divided up. First is the two sets of 7, of which Biblical *Revelations* is famous for. This makes one central line, the diameter. The 7 and the 7 as a pair of numbers is mentioned over and over again, or just the 14, or the 7 itself alone. All are used to encode the 14-fold division with the $1/7^{th}$ and $2/7^{ths}$ angles.

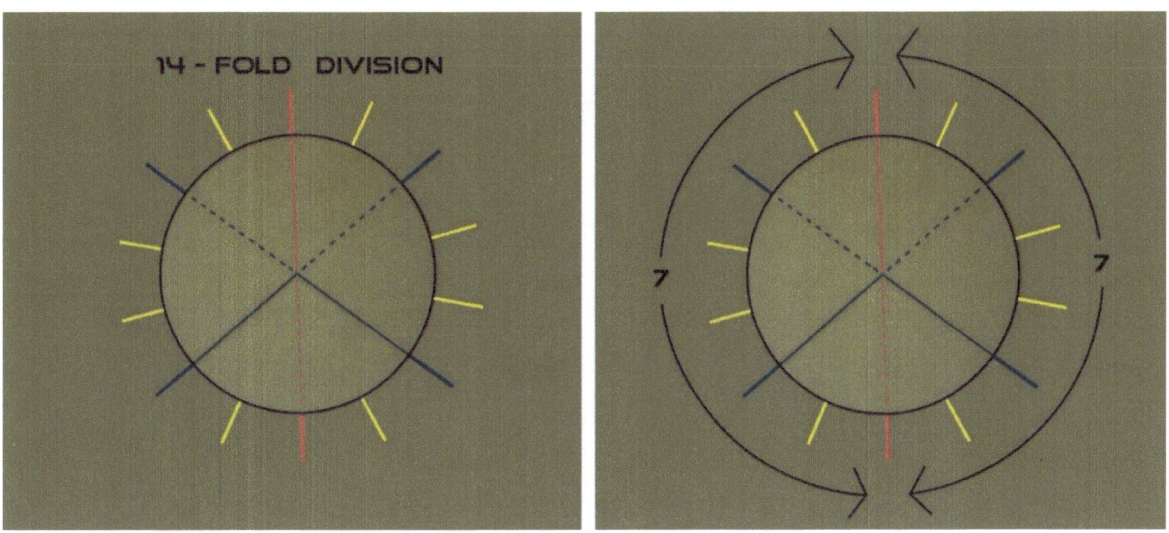

Second, there are two intersecting lines that create a 4-3-4-3 'X' shape, with their specific angles. And third, the 14-fold division also holds the specific triangle shape of 5 parts, 4 parts, and 5 parts, the 5-4-5 trine.

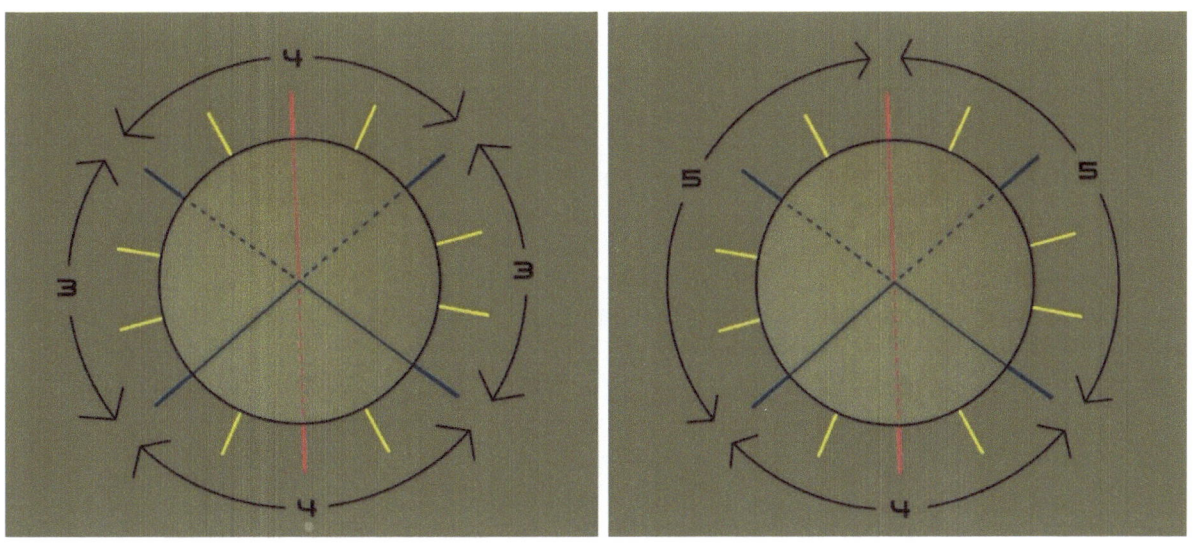

Chapter 25: The Circled-Square & Cubed-Sphere

1/7th and 2/7ths... or 2/14ths and 4/14ths, are the measures that are used to create the Circled-Square and the Cubed-Sphere of sacred geometry, For the ancient mathematicians, the Circled-Square and the Cubed-Sphere were considered PERFECTION IN 2 DIMENSIONS AND 3 DIMENSIONS.

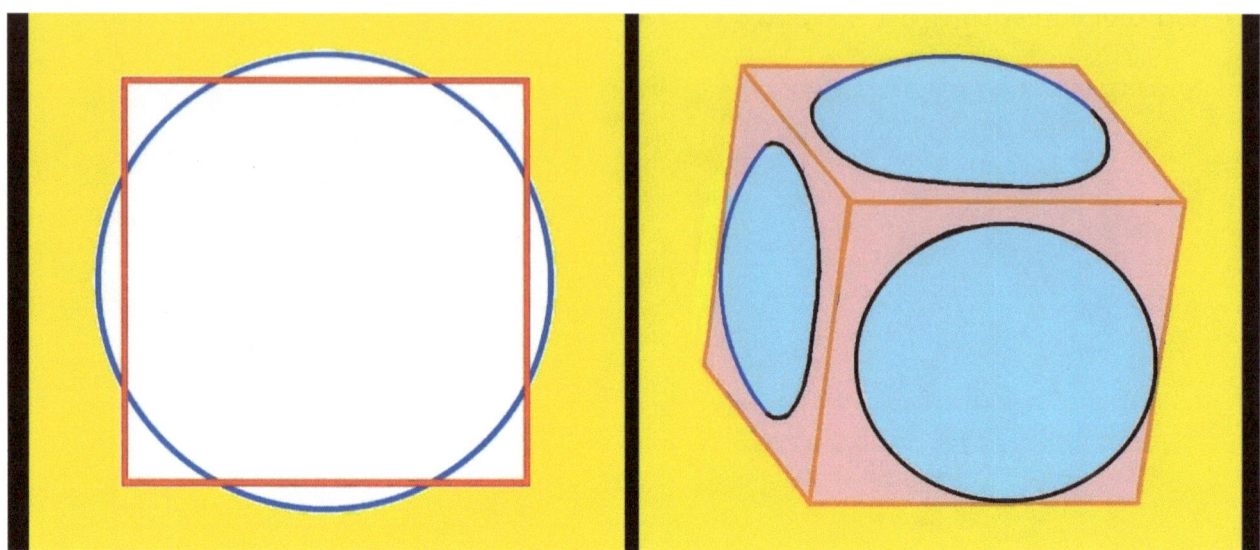

the Masonic Checkerboard Cross holds the angle of 1/7th – 2/14ths, very closely. The true division for 1/7th of a 360 degree circle is 51 and 3/7 or 51.42857143 degrees. And the 27x13 rectangle of the Freemasons is very close with 51.4199076. The difference between them is only...0.00866383, and that's very close.

The 27/13 is almost so perfect, just by itself; in its accuracy to within 1/800 of true 1/7. And it's also got the 27, which is a factor of 108. And 108 is the all-important number in Archeo-Astronomy, represented by clues like the 108 stiches in the Masonic apron.

And now I have set the stage to unveil one of the greatest mathematical secrets of the ages. I believe this is part of what is called the Masonic 'lost lore'.

A cross made from 2 Masonic checkerboards can be used with the 2/7ths – 4/14ths measures; to create one of the two ratios for both the Circled-Square and the Cubed-Sphere in the 2nd and 3rd dimension. The perimeter of the Circled-Square and the volume of the Cubed-Sphere.

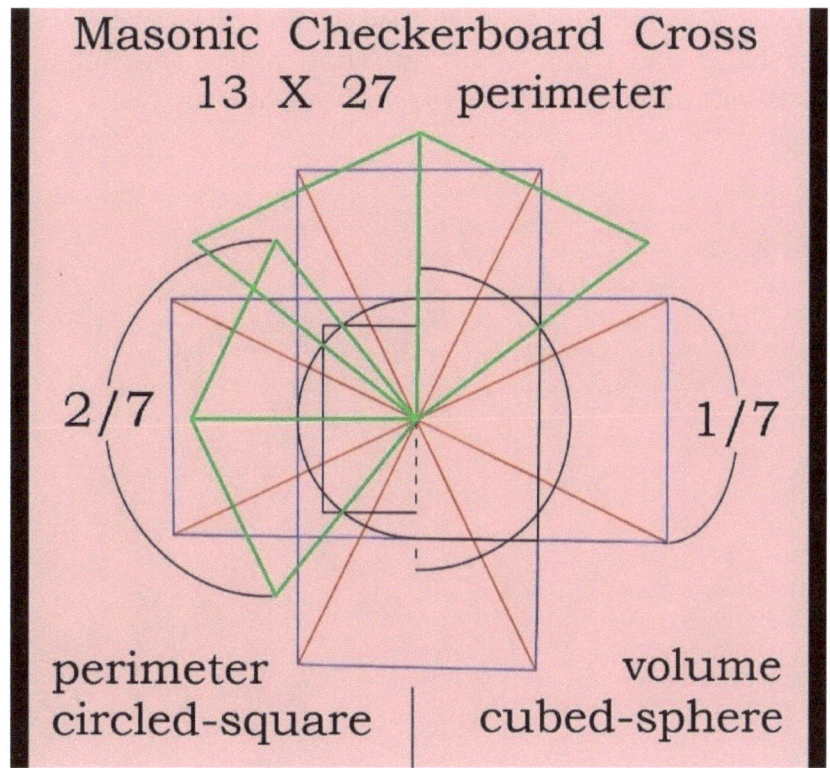

Chapter 26: The Beast & Dragon with the 5-4-5 Triangle

This measure of 5-4-5 was encoded in Biblical *Revelations*, with one of the three great serpents in the heavens, Draco itself. The constellation, Draco, is centered around the North Ecliptic Pole, and the constellation roughly marks the Circle of Precession. This is the round path that the North Celestial Pole takes in the heavens for 25,920 years. Also, Polaris, the current North Celestial Pole star is very close to Draco. As well, the more ancient constellation of Draco had wings; of which our current North Pole Star, Polaris, was part of.

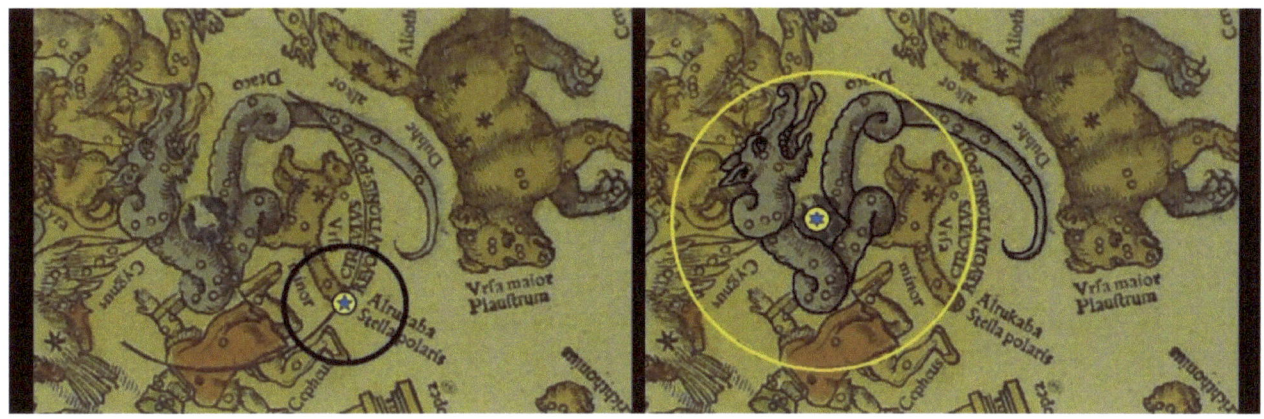

This is why in *Revelations*, both the Beast and the Dragon have 7 heads and 10 horns. If each head has a pair of horns. That means the Beast and the Dragon each have 5 heads with horns and 2 heads without horns.

When you add the Beast and the Dragon together that creates; 5 heads with horns, 4 heads without horns and 5 heads with horns. This creates the 5-4-5 split of the 14-fold division. The $2/7^{ths}$ angle is made from the 4 heads, 2 each from both the Dragon and the Beast that have no horns. And when you tally up the heads with crowns it makes the $1/7^{th}$ angle; as the Beast alone has 2 heads with no crown. This 5-4-5 Trine, and also it's $4/14^{ths}$ angle, is the pyramid in the famous symbol of the Eye and the Pyramid. Did St. John, the Divine Revelator, truly see these in a vision, like he did the Leonids Superstorm Meteor Shower? Or did he create the image consciously to encode the 5-4-5 trine? That is a true mystery for the ages.

Chapter 27: The Giza Pyramids of Egypt

All of the pyramids of Egypt are said to have a specific angle from the apex to the base It is roughly a 51.42 degree measurement, a 1/7th angle.

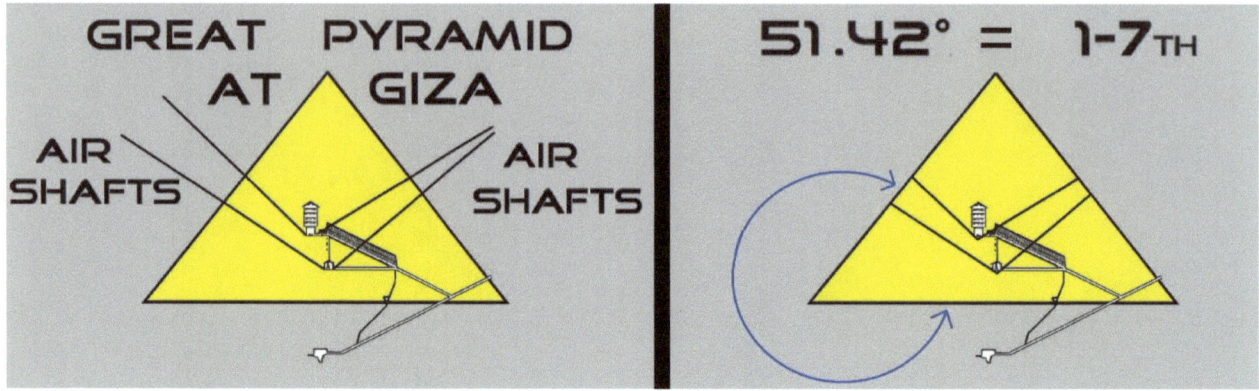

In *The Message of the Sphinx*, (by Graham Hancock, Robert Bauval), it has a diagram (p.52; TMotS) that possibly shows the Great Pyramid of Giza also encodes the 4-3-4-3 'X' and the 5-4-5 triangle. In the Great Pyramid of Giza there are the famous, four hollow shafts. One pair 'V'-ing upwards from the king's chamber, and the second pair ascending from the queen's chamber. It seems to me that they both hold the measure of 2/7ths, or 4/14ths, which is all important, for both the Cubed-Sphere and the Circled-Square.

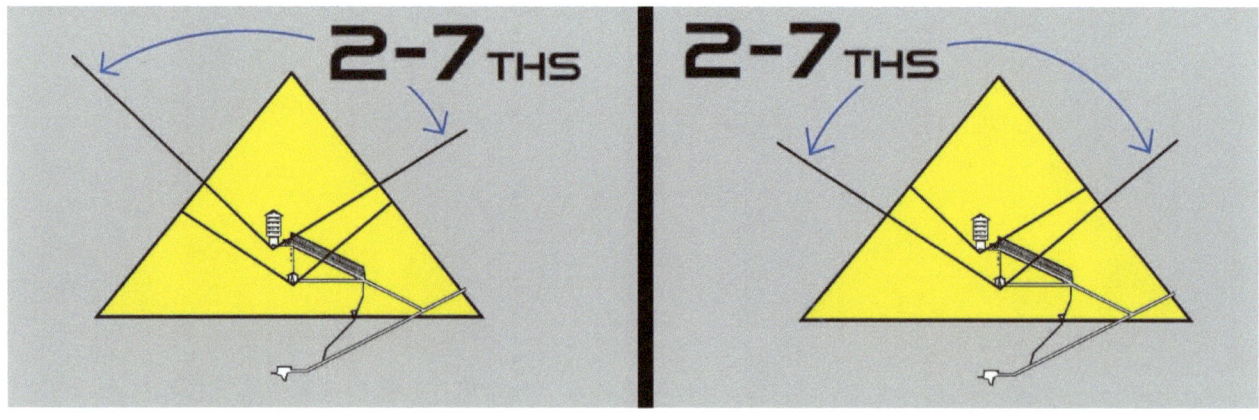

Chapter 28: The Pyramids at Chichen Itza

At Chichen Itza, in Mexico, is the famous Toltec/Mayan Pyramid of Kukulcan. It is famous with its sunset display for the serpent of light and shadow running along the west edge of the north stairs. This pyramid holds both, the angles of the sevenths and the fifths; which are each fundamental parts of sacred geometry. As can be seen, for Kukulcan's pyramid, over and over from different ground points of perspective.

Also at Chichen Itza is the lesser famous, Pyramid of the High Priest's Tomb. And while the more famous pyramid of Kukulcan had to have the angle of its stairway specifically slanted; to achieve the phenomenal serpent of light and shadow (for moonlight as well as sunlight); the High Priest's Pyramid was under no such limitations. Notice how each side stairway slopes out from the top at $1/7^{th}$, making the total width at $2/7^{ths}$ or $4/14^{ths}$.

And this holds the all-important 5-4-5 split of the 14-fold division, and of course the 4-3-4-3 'X' shape. What's important is the, 4 parts out of 14, which is made from both of these divisions of 14.

....

Chapter 29: Copernicus, Chichen Itza, Stonehenge, & Washington D.C.

What's truly fascinating to me is that I've seen proof left in art that this 5-4-5 triangle was known to others. For instance, there is Copernicus the astronomer who said the Sun was the center of the solar system. Here is a famous page of his own writing and his famous drawing.

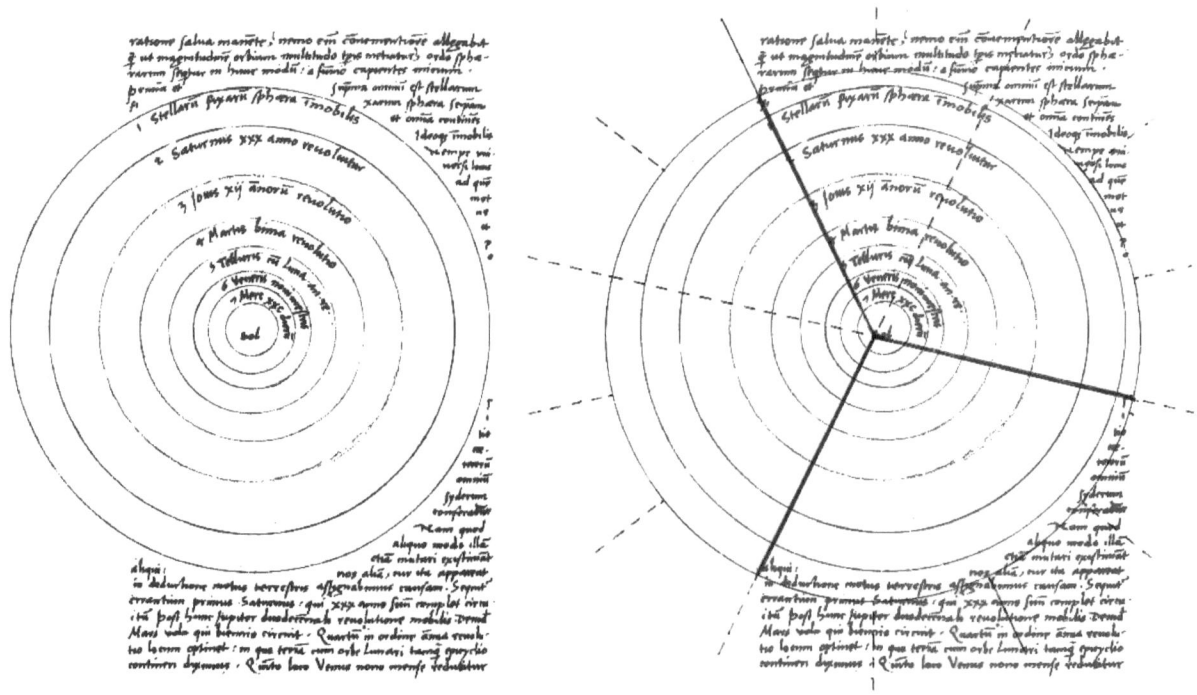

And there is a famous artist of Astronomy, Andreas Cellarius, who may have been guided by Copernicus, or he already knew about the 5-4-5 trine.

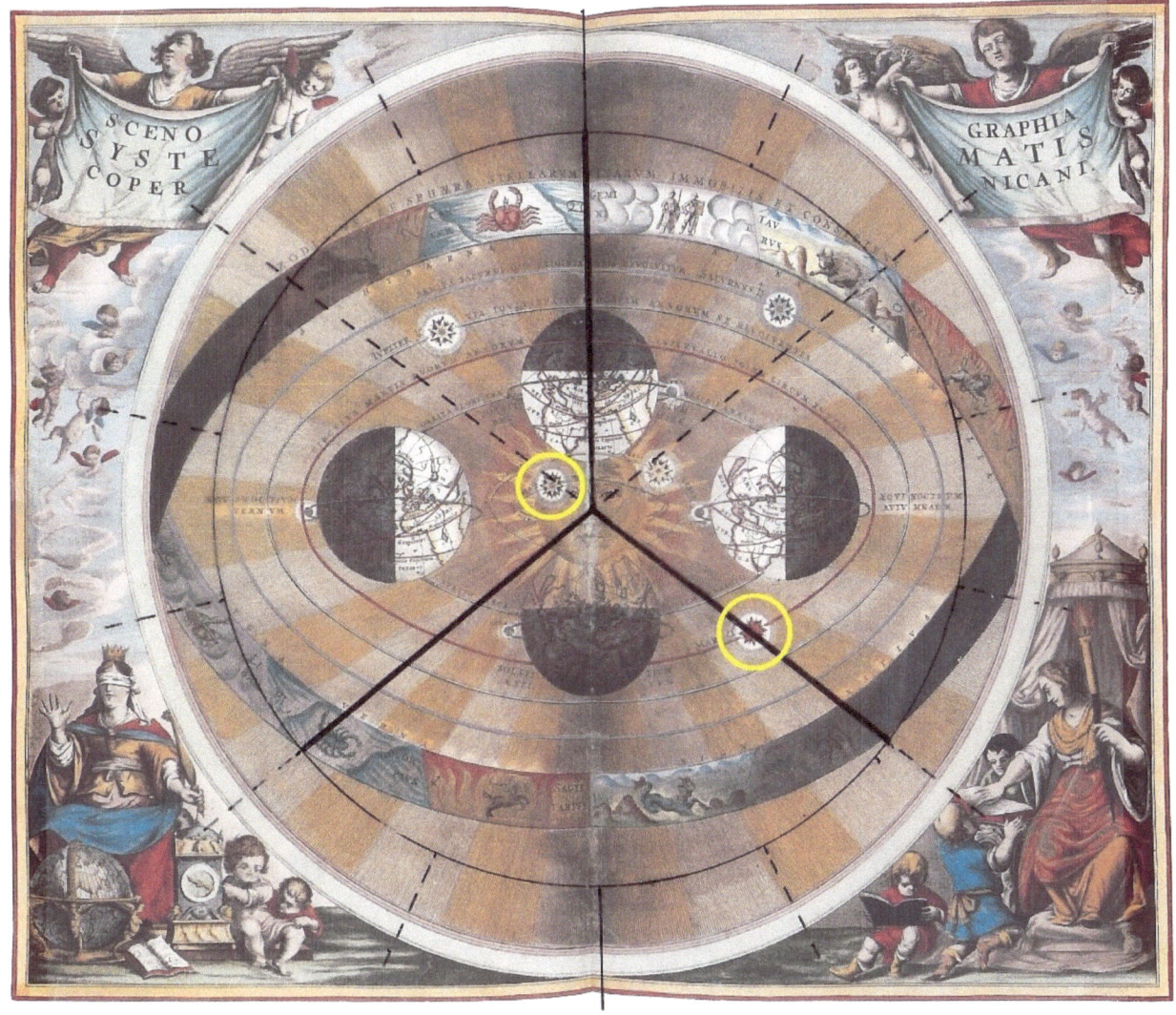

The latitude of Chichen Itza, is mapped out for the extreme sunrise and sunset positions of the Dec. and June Solstices along the horizon. And this is incorporated into the pyramid's design. There is an incredible book, a pamphlet almost, by Lic. Miguel Angel Vergara C.; it's called *Chichen Itza; Astronomical Light and Shadow Phenomena of the Great Pyramid.* And while it's not stated in the book itself, when its diagram is examined it holds the 1/7th sacred measure. Just as it does in the Masonic checkerboard with its complete 27 by 13 square rectangle (12 by 26 square tiles... and the half-tile width encompassing border). This sacred measure of 1/7th, marks the extreme north and south between the sunrise and sunset positions of the two solstices. This solar orbit phenomenon for the 1/7th, 51.42 degrees occurs at; 20 degrees, 40 minutes, of latitude; whether one is north or south of the equator. And nowhere else.

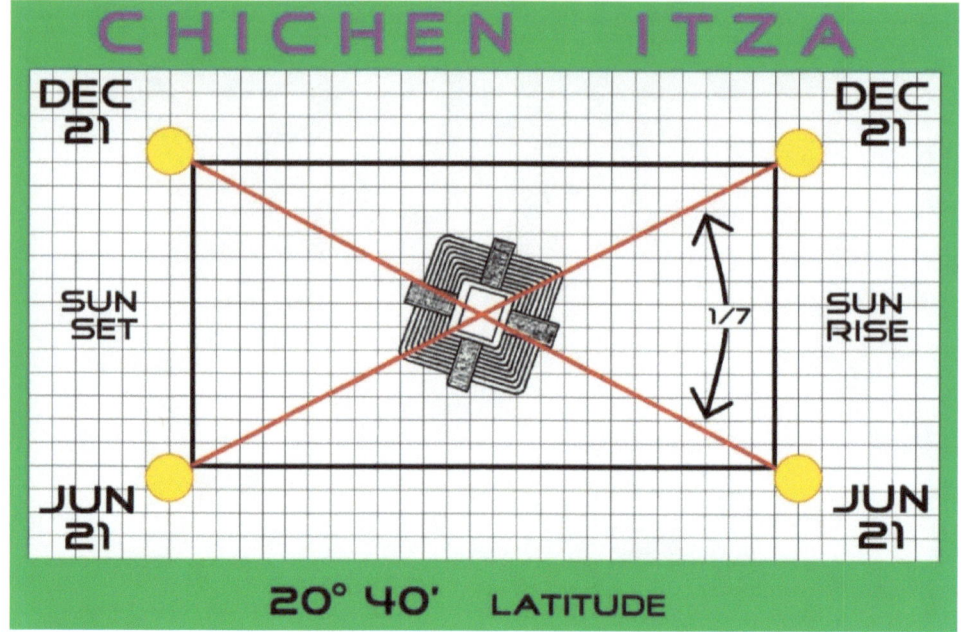

And this is the ground plan of the main site of Chichen Itza.

Map data: Google, DigitalGlobe, 2014

It is very apparent that the Toltec/Mayan architects of Chichen Itza encoded the knowledge of the 5-4-5 spilt of 14 parts into the ground design. There are two pathways on the site that have 5/14^{ths}, and 1 pathway that has 4/14^{ths}, just like the 5-4-5 division of 14. The second path with the 5/14^{ths} may be original or from the 1922-1942 reconstructions by the Free Masons.

Map data: Google, DigitalGlobe, 2014

Map data: Google, DigitalGlobe, 2014

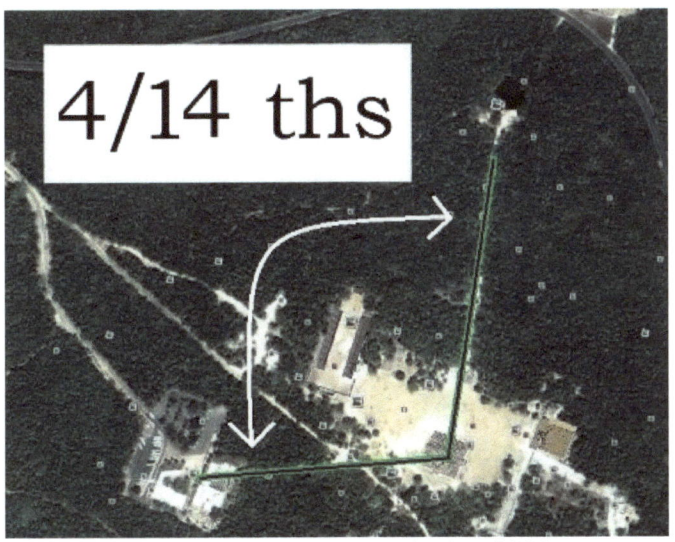
Map data: Google, DigitalGlobe, 2014

In sacred geometry; the shape of the five-pointed star, and the pentagon with its $1/5^{th}$ division; is known to hold the divine proportion.

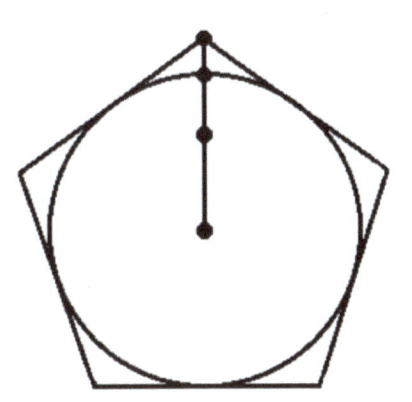

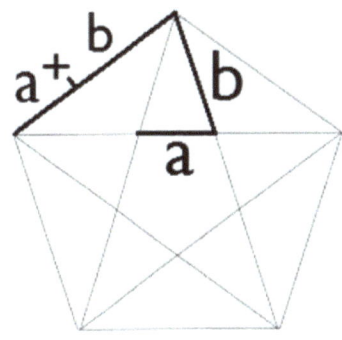

And since Chichen Itza is in the tropical latitude it gets a zenith sun passage, when the sun crosses the direct center of the sky. And it just so happens, that the sunset of the zenith passage; which is marked by the Upper Temple of the Jaguar, on the famous Great Ball Court of Chichen Itza; is at 1/5th, 72 degrees, west of due North. So the latitude of Chichen Itza at 20* 40' holds the 1/7th angle between the solstice sunrise and sunset along the horizon, and it also holds the 1/5th angle down from true north, for the zenith sunset.

Map data: Google, DigitalGlobe, 2014   Map data: Google, DigitalGlobe, 2014

....

This specific 51.42 degree measure itself was also used as latitude markings for two very important cultural and historical sites. Stonehenge has its latitude at 51.17 degrees, up from the equator. And Washington DC is at the latitude 38*53'42" up from the equator; which puts it 51*06'18" down from the Celestial North Pole. I believe that the 1/7th measure specifically used to place both Stonehenge and Washington DC. More about the importance of the latitude at these sites can be found in the books; *Uriel's Machine* (Robert Lomas & Christopher Knight), and *The Secret Architecture of our Nation's Capital* (David Ovason).

....

Chapter 30: The Galactic Axis & Ecliptic Axis

During precession, the Celestial Axis takes its swiveling, circular path around the Ecliptic Axis, on what's called the North and South Circles of Precession. But there is also the Galactic Axis. As these three Axis' affecting the Earth are seen from a side view, if you will; it shows all three axis' in their lengths. So, as well as the swiveling Celestial Axis, there are the two relatively unmoving; Ecliptic Axis and Galactic Axis, which are separated by 60 degrees.

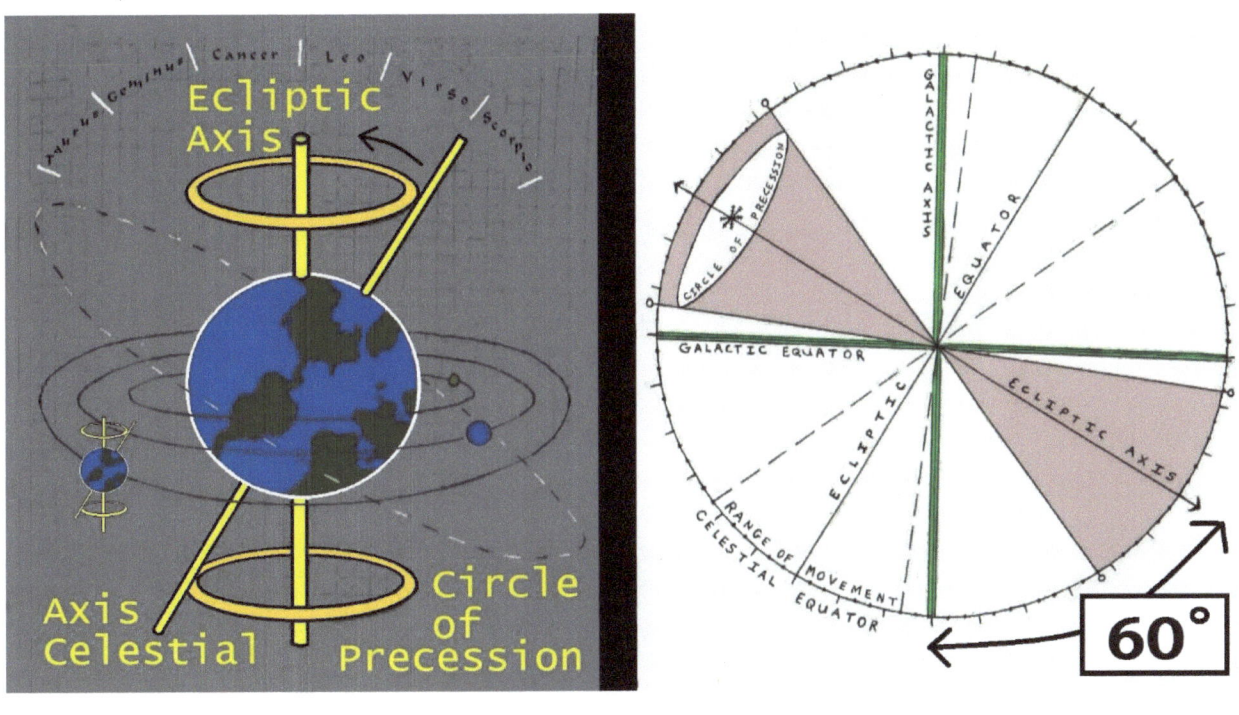

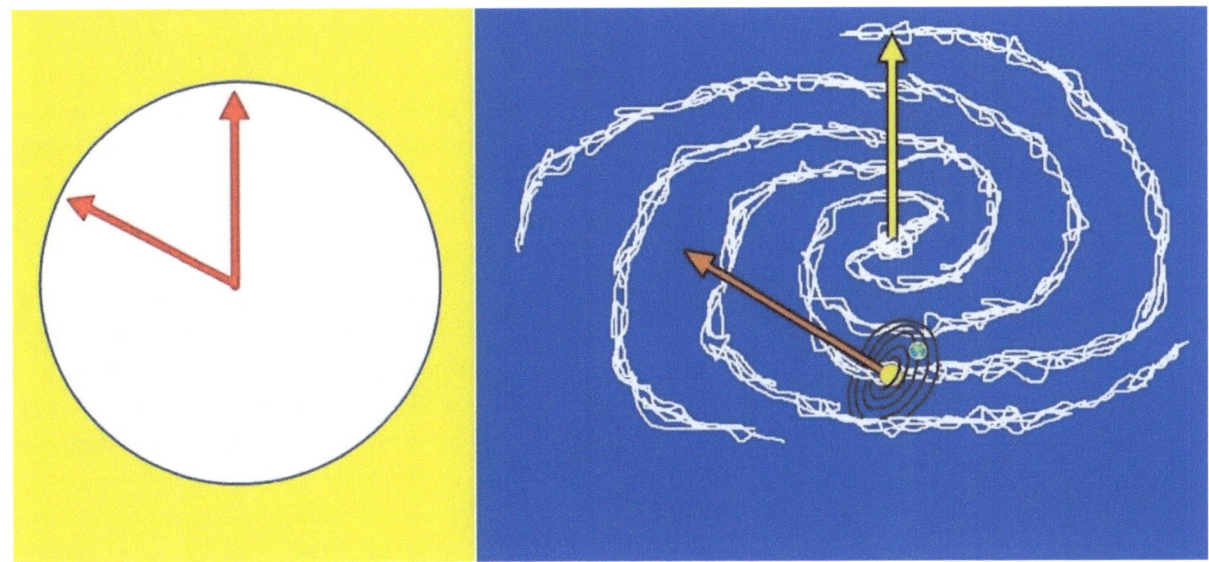

UNDERSTANDING REVELATIONS BY ASTRONOMY

And by my calculations, it's as if they were the hands of a clock, sharing the plane of the clock's face, with one hand at 10 and the other hand at 12.

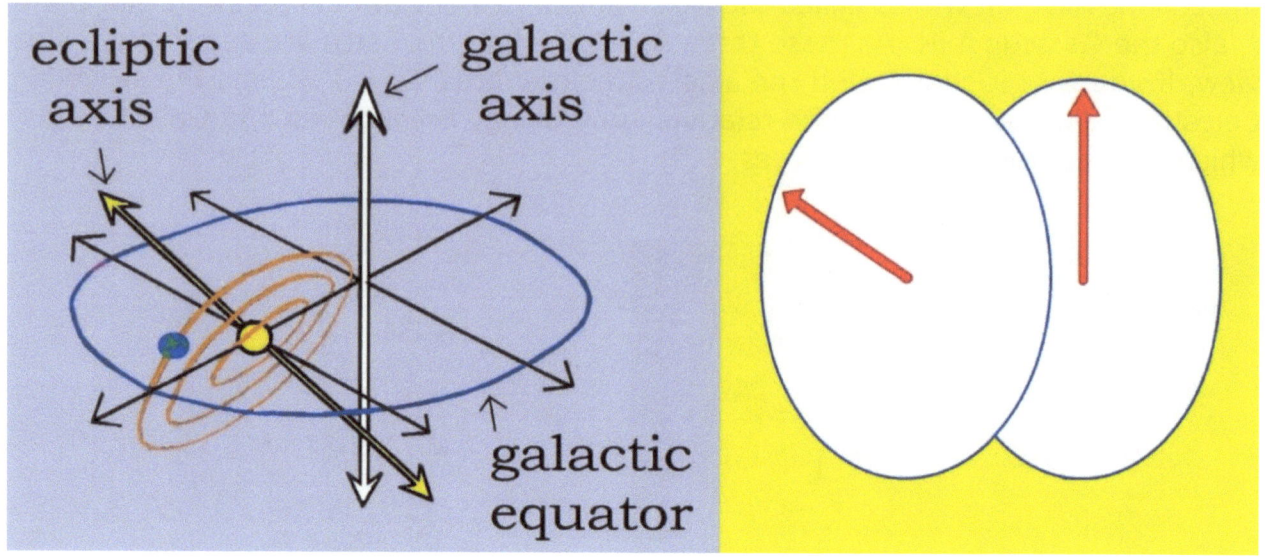

Chapter 31: Sighted Down the Earth's Celestial Pole

There is another important perspective to look at these 3 axis'. It's with our eye line always centered on the moving Celestial Axis, sighting down its length; as if we are at the South Pole, looking straight to the North Pole.

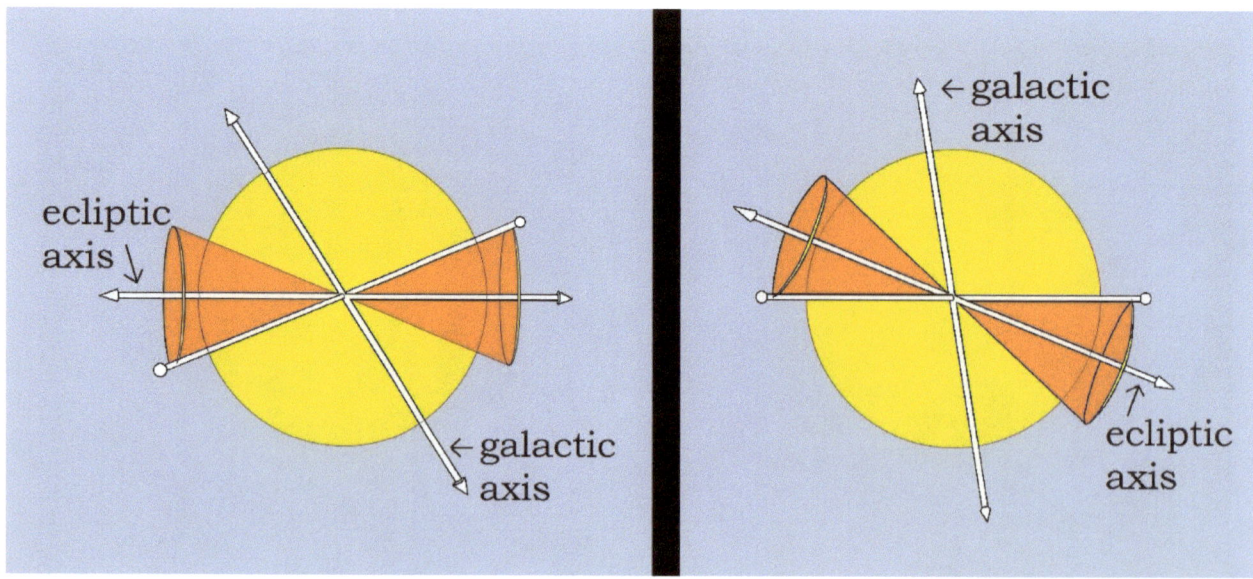

This moving perspective makes the Celestial Axis change from a line to a dot. And when the other two axis', the Ecliptic, and Galactic, are looked at from this perspective; it creates an optical illusion. First, that the Celestial Axis, centered in the diagram, stays in one place as the dot. And second, that the Galactic Axis appears to move around and around like a propeller blade; with its center running through the central dot of the Celestial Axis. And finally, the Ecliptic Axis also stays in one place, seen in length as an axis. In the diagrams, I have included the Circles of Precession, with the circled-cross of the Solstices and Equinoxes inside them, matching the diagram above.

Sited down the Celestial Axis

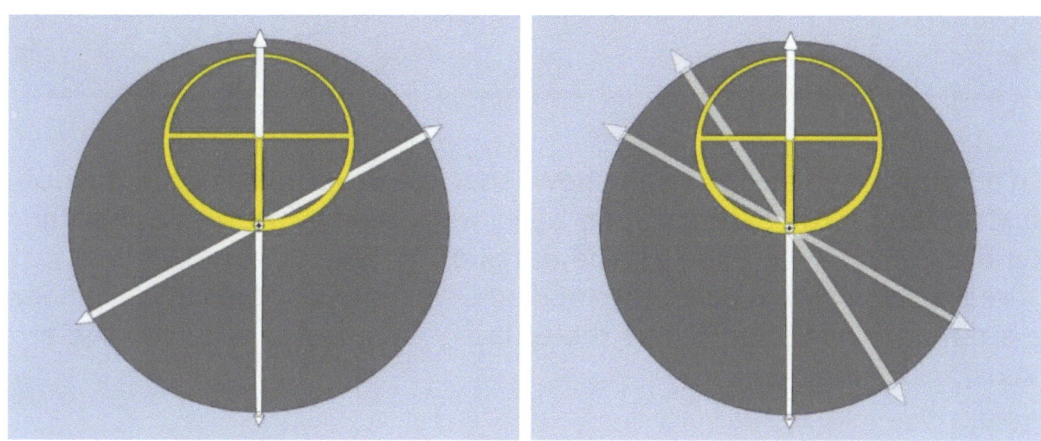

This means that twice during Precession, the Galactic Axis and the Ecliptic Axis will appear to merge as one. The illusion shows the Galactic Axis moving in and out of conjunction with the stable Ecliptic Axis. And this also means that twice during precession, the Galactic Axis will move perpendicular to the Ecliptic Axis; which creates the equal-laterally divided, cross shape.

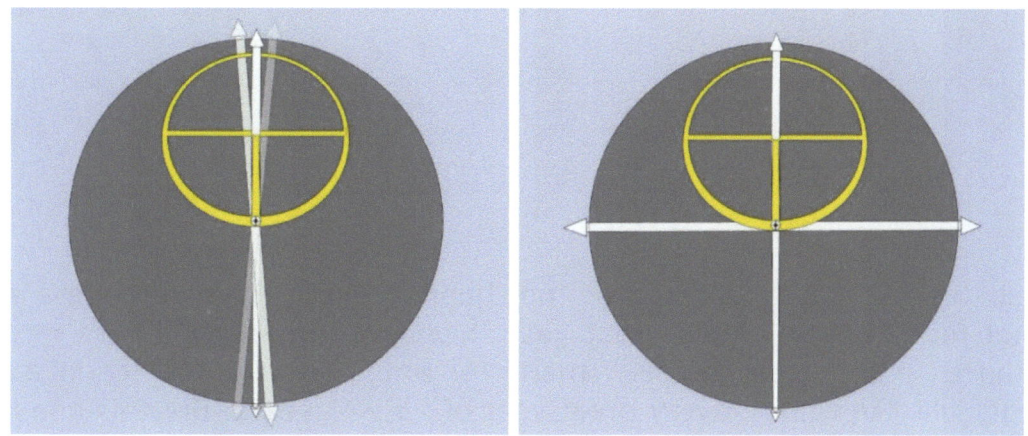

But what also happens, most importantly; is that four times during precession that cross shape is not equilateral, it's an 'X' shape, specifically in the 4-3-4-3 division of 14 parts. And right now in the Great Year's cycle of Precession, as we are at the Galactic Alignment's stop and start point for the entire 25,920 years; the Galactic Axis and the Ecliptic Axis are in the 4-3-4-3 'X' shape.

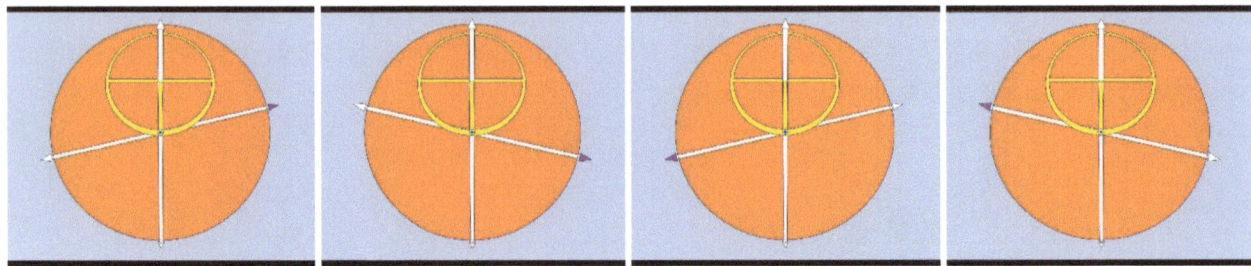

On a hemisphere star chart, it shows that the Ecliptic Axis is on the north/south Celestial Meridian for the dates of June 21st and December 21st; and that the Galactic Axis is on the Celestial Meridian for the dates of October 2nd and April 3rd. From this illusion, created by sighting down the moving Celestial Axis, the constellations appear to move around in a wobble; just as the Galactic Axis 'appears' to revolve around the Ecliptic Axis.

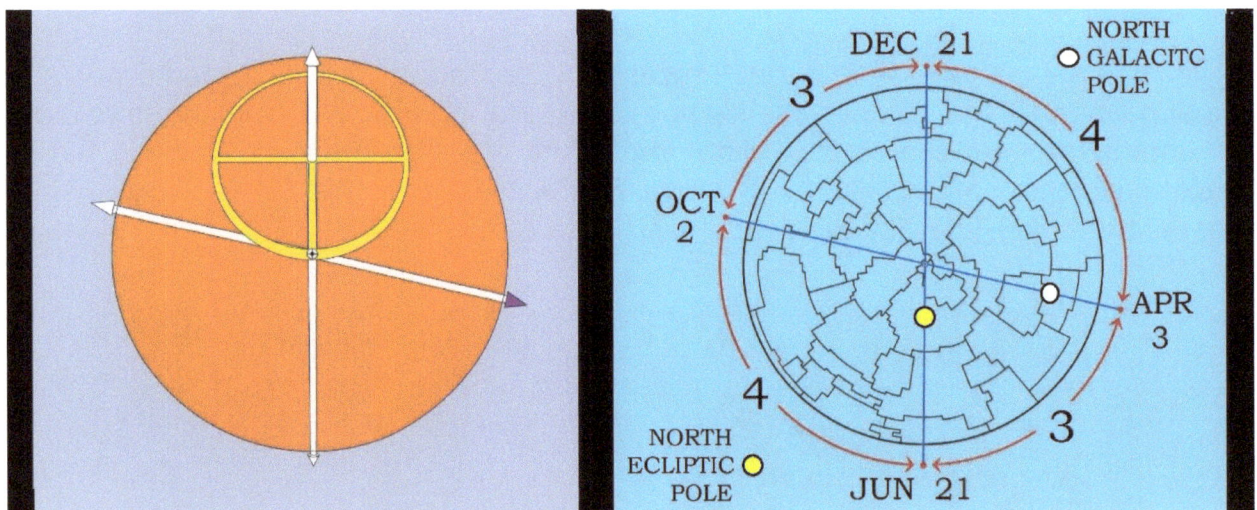

Yet this 4-3-4-3 'X' shape does not happen every time we have a stop and start point for the cycle of the great year. I believe we are in a very rare, set of circumstances brought by the slow march of time. So yes, we are in a Galactic Alignment, just like we are every great year of 25,920 years. But this one has a lot more than just the stop and start in the cycle of the 'Great Year'.

....

UNDERSTANDING REVELATIONS BY ASTRONOMY

Chapter 32: Once Again, the Sacred Geometry

For many of the ancients, who appreciated the divine harmony of mathematics; there is nothing more perfect in the Godhead's orchestration of reality, than the Cubed-Sphere and the Circled-Square, of the third and second dimensions.

As well; both dimensions each, have their own two formulas.

In the Circled-Square, first there is the measurement of the two perimeters as equal: length x 4 = radius x pi.

Second for the Circled-Square is the measure of the interior areas as equal: length squared = radius squared x pi. It is almost 8 units and 9 units wide.

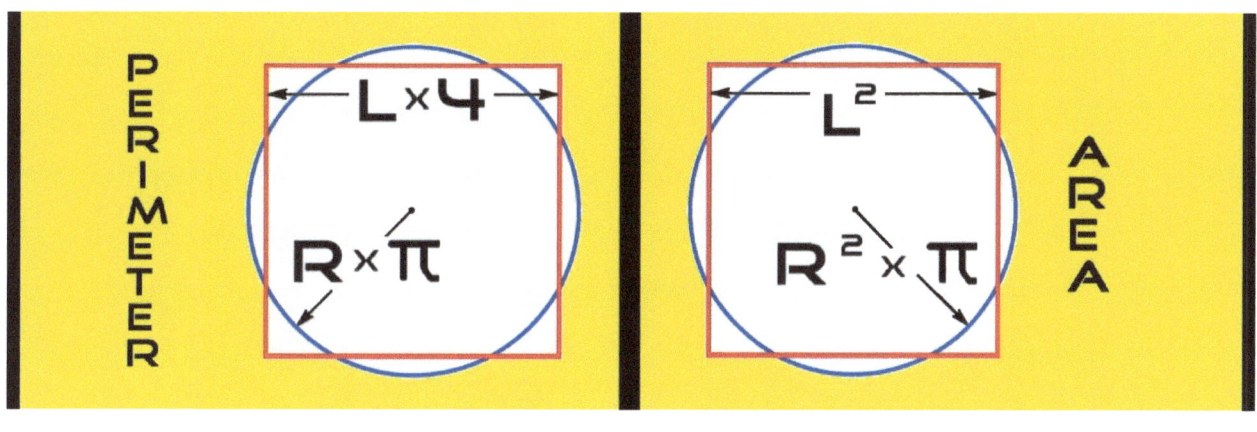

....

For the Cubed-Sphere, first there is the measure of the surface areas: length squared x 6 = 4 x pi x radius squared. The earth to moon ratio.

And second for the Cubed-Sphere is the measure of the interior volumes: length cubed = 4/3 x pi x radius cubed.

The shape for the pyramids of Egypt is famous for holding the pattern of equal volume for the Cubed-Sphere.

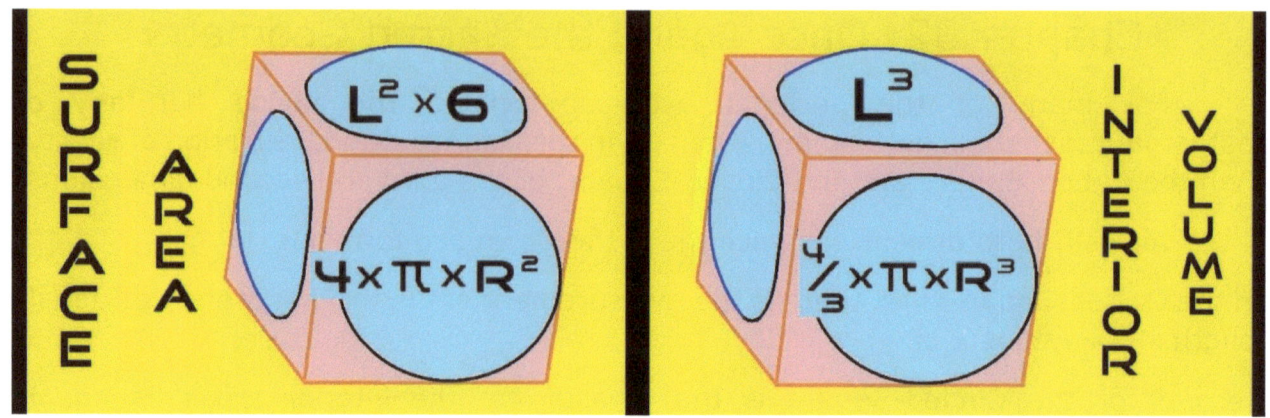

To bring it full circle. In my opinion, for many of the advanced encodings of the mythos it's not the 7 itself that counts, it's the 14, and more specifically the two sets of 7 that was considered important. As I stated, it creates the 4-3-4-3 'X' shape; and the 5-4-5 triangle. Between these two measures used to divide the 14-fold division, it was the 5-4-5 split that was most prominently encoding in Biblical *Revelations*. As I have shown, it relates to the angles as the divine measurement, between the Ecliptic and Galactic Equators. And, there is more.

Chapter 33: The 5-4-5 Split of the Calendar

I believe this 5-4-5 split was applied to the calendar of the solar year. In *Revelations 12*, among other places in St. John's text, it mentions specific blocks of times; in years, months, and weeks. The full total of which, in days, is said to be 1,260 days. This continuous repeating of the same length of time in days, weeks, months and years, suggests a calendar to me. And also the length of days is put at 260 plus 1000.

Yes, different cultures are different cultures; but in the end, math, is math.

In the calendar of the solar orbit, there are 365.2422 days. And when dividing that tropical solar year with the 5-4-5 split, this creates the 260-day sacred calendar of the Meso-American cultures. If the 260-day calendar aligns with the 364-day sacred calendar, this creates an exact measurement of 14^{ths}. The mathematical ancients all over the world would surely have been aware of this. For synchronizing the 5-4-5 triangle within the solar year as I have, I am using the placement given by John Major Jenkins in *Maya Cosmogenesis 2012*. The 260-day calendar starts on August eleventh or twelfth; which, for me, holds the extra 'day outside of time', between the 364-day sacred calendar and the 365-day secular calendar.

When starting the 260-day calendar at this date, the midpoint at 130 days in this 260-day Tzolkin Calendar is the all-important December Solstice. The end of its count is reached after another 130 days is on May first. The 260-day calendar is exactly 10 parts of the 14-fold division inside of the 364-day calendar, which makes it the 5 parts and 5 parts, or two sets of 130 days. With the 260-day calendar inside of the 364 or 365-day calendar it makes the the 5 & 5 of the trine. This leaves an additional 104-105 days, and it is this portion that amounts to the last 4 parts of the 14-fold division.

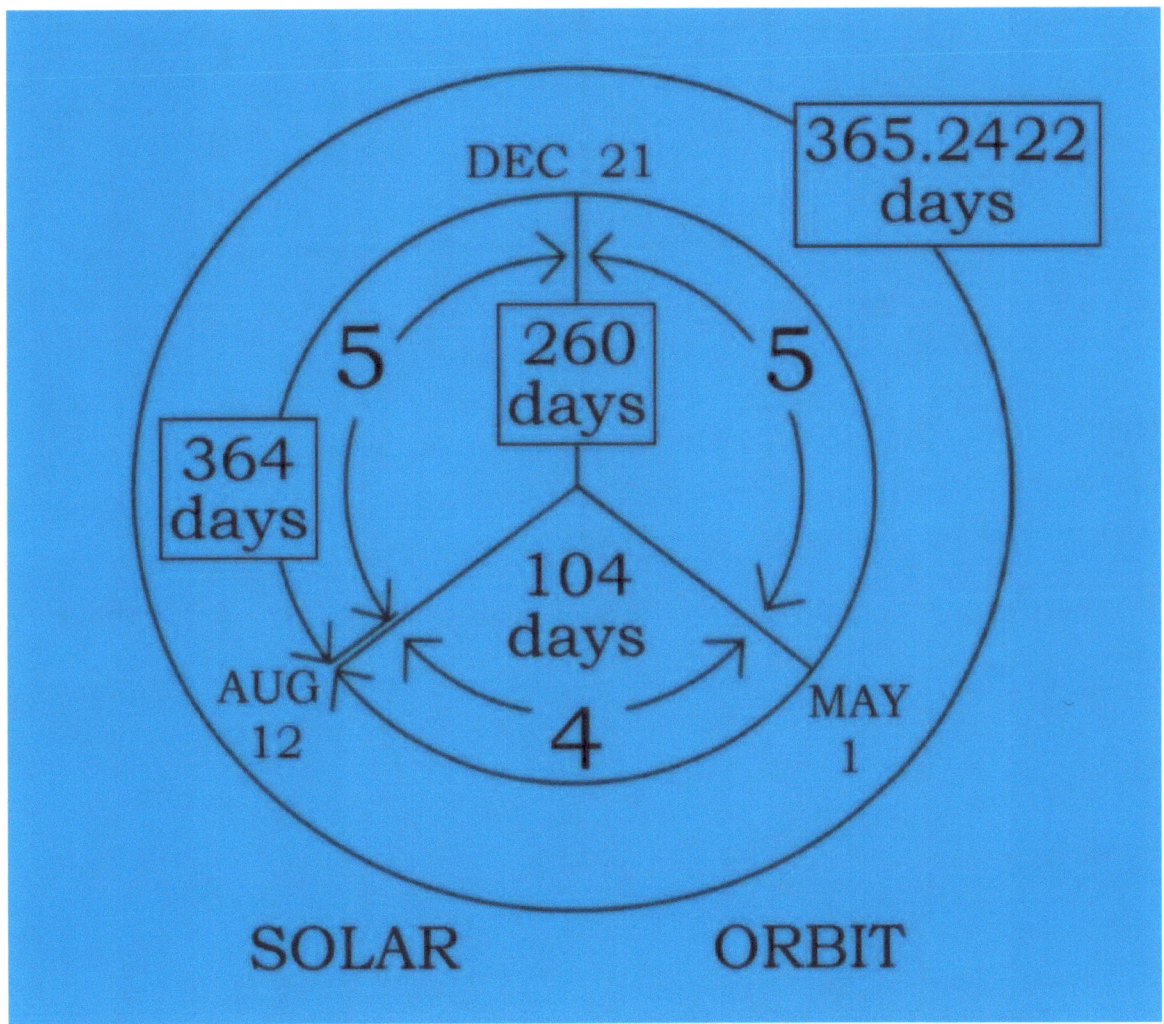

Genesis 5:3 tells us that Adam (and Eve also?) lived 130 years before he begat Seth. And in *Genesis 5:5* it says that; Seth, himself, lived 105 years before he begat his son, Enos. Which is 365 years, encoded for the calendar days. Clearly it shows the 5-4-5 trine of 14 parts.

....

Chapter 34: The 4-3-4-3 'X' shape & the Calendar

Every year holds this union of the 260-day calendar and the 364/5-day calendar. In Precession here and now, the Ecliptic and Galactic Axis create the 4-3-4-3 'X' shape. The Solstices of Dec. 21 and June 21 have the Ecliptic Axis on the north/south Celestial Meridian at noon and midnight. And by my best calculations Oct. 2 and April 3 have the Galactic Axis on the Celestial Meridian.

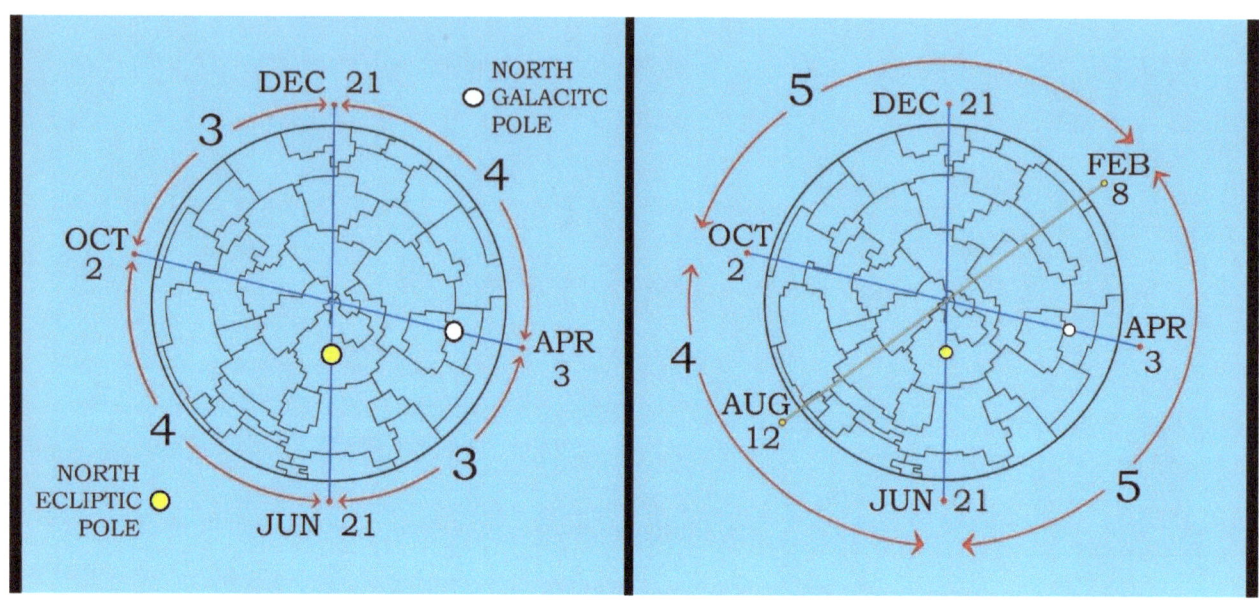

There is one more division of the solar year using the 4-3-4-3 X-shape.

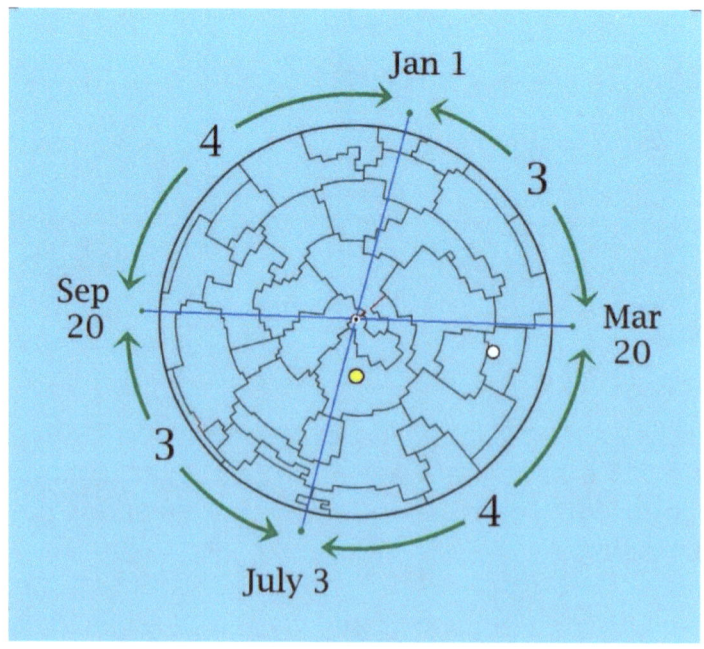

UNDERSTANDING REVELATIONS BY ASTRONOMY

This 4-3-4-3 'X' has one axis through the Equinoxes. And it connects to the other axis going through the dates of Jan. 1st and Jul. 3rd. The Equinoxes and the Solstices are bound to the calendar of the solar year, so those events do not move with Precession. But for the Jan. 1st and Jul. 3rd Axis there are events which do indeed move through the calendar dates with Precession. And it's not just one stellar event. Right now there are several sets of alignments on, and centered around, the dates of January 1st and July 3rd.

....

PART III – The Greater Alignments of Precession
Chapter 35: A review so far…

Let's review about the very special alignment that we are in right now.

Many people are aware of the Precession of the Solstices and Equinoxes, and its cycle just under 26,000 years. But not every Great Year and its cycle of Precession holds all the alignments that affect the Earth right now. To begin with; there are three Axis that spin, and create their planes of centrifugal force, called equators. First is the daily Earth spin, named the Celestial. The second is the Earth's annual solar orbit, named the Ecliptic. And the third is how the Earth is positioned in our Milky Way, named the Galactic.

These three, each by themselves, is easy enough to understand, but throughout the Great Year and it's 25,920-year cycle of Precession, this set of three gravitational forces interact in a series of complex ways.

Both the Ecliptic and the Celestial have an equator and an axis. The two share the same space of the sphere and are slightly offset from each other by almost 23 and 1/3 degrees. They intersect at the two Equinoxes and are farthest apart at the two Solstices. There is one colure / ring connecting the points of the North and South Celestial Poles with the two Equinoxes. And there is another Colure that connects the North and South Ecliptic Poles with the two Equinoxes. These are called the Ecliptic and Celestial Equinox Colures. (see diagrams on pages 9-12) There is one more colure that is shared by the both the Ecliptic and Celestial Axis' with their Poles; and it is also shared by the Solstices points for each of their Equators, one ring that binds them all together. The Solstice Colure.

I believe that, for all these stellar events, what was encoded most into the ancient, world-wide mythos; was how our alignments now, at the stop and start point of the great year; create 1 of the 2 formulas of the 'Circled Square', and 1 of the 2 formulas of the 'Cubed Sphere'. These were what many ancient cultures with the higher mathematical skills believed were the most perfect expressions of divine order in our reality.

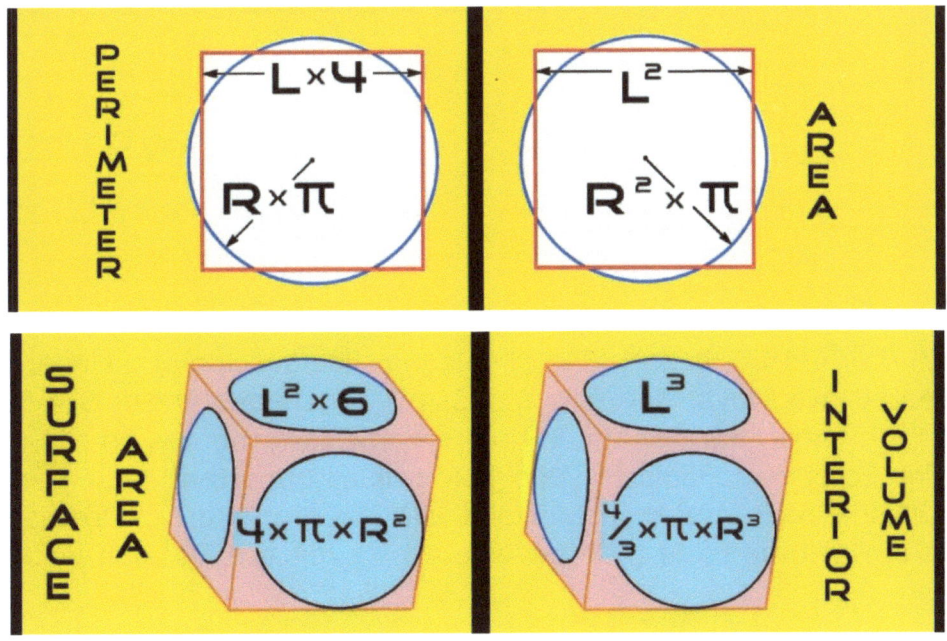

UNDERSTANDING REVELATIONS BY ASTRONOMY

These 2 formulas, one for both the 2nd and 3rd dimensions, are obtained by using a very thick, equal-armed cross. It is made from the famous Masonic checkerboard; which I have turned into the 'Masonic Cross'. I place the 2-7ths angles over the Masonic Cross making the Cubed-Sphere and Circled-Square.

This diagram shows how to create the perimeter of the Circled-Square, and the volume of the Cubed-Sphere from the Masonic Cross.

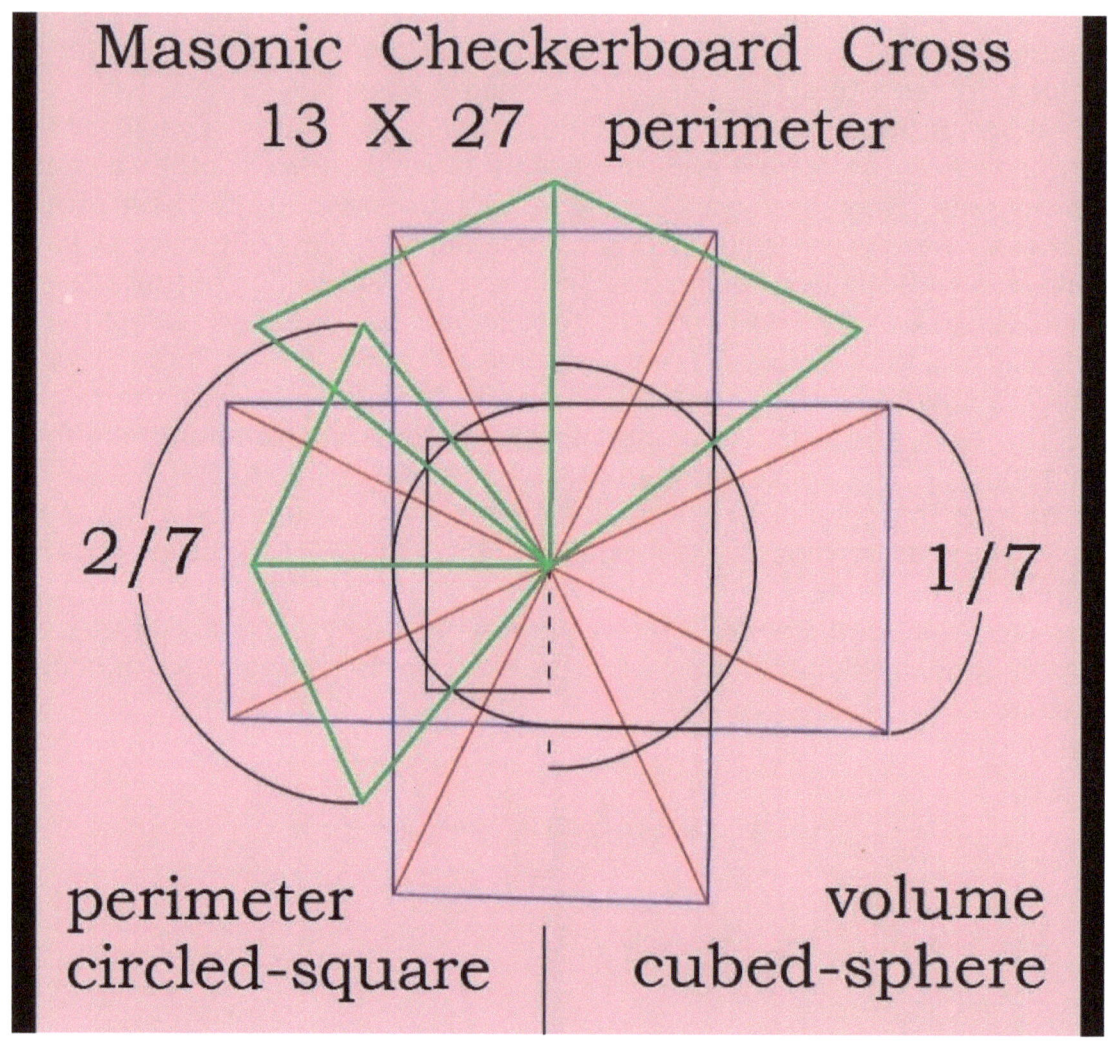

The 2-7ths / 4-14ths, is part of a 14-fold pie slice division of a circle. This 4-14ths angle is created two ways. There is a 4-3-4-3 'X' shape. And a 5-4-5 three-way division, like a triangle, a trine. And in the here and now there is an alignment of the two Axis'. The Ecliptic and Galactic Axis' hold this all important 4-14ths angle sighted down the Celestial Axis. And this coincides with the stop/start of Precession. It was all part of a stellar time table to know when the expected Messiah would arrive, in our Era.

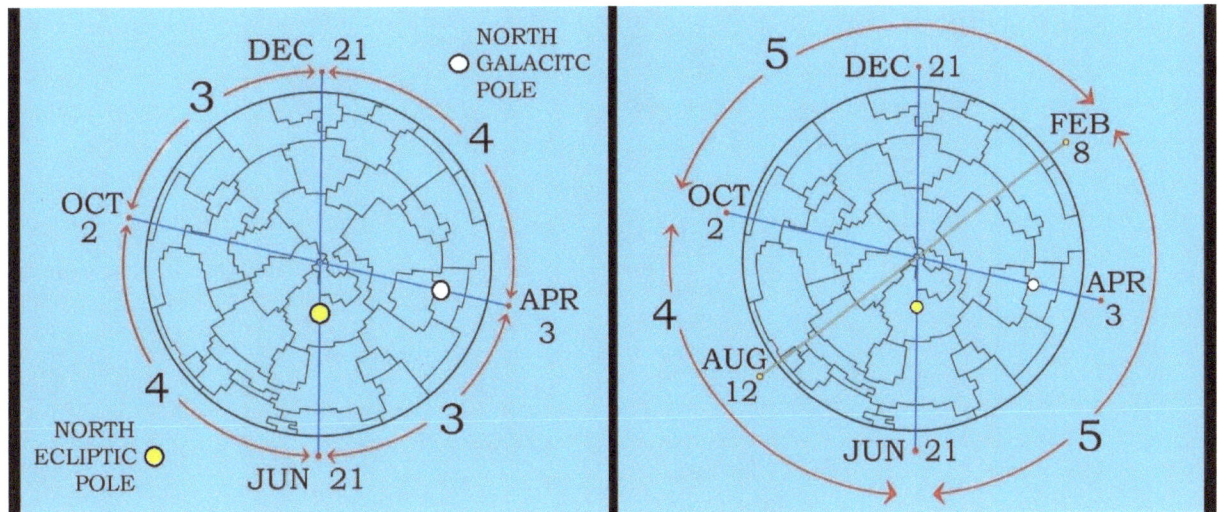

And this was encoded in many of the worlds mythos and religions, as with the *Revelations* of Saint John the Divine. With the Dragon and Beast's 14 heads, the heads without horns is 2-7ths and the heads without crowns is 1/7th.

BUT THIS POSITION OF THE 3 AXIS, THE CELESTIAL, ECLIPTIC AND THE GALACTIC, CREATING THIS 4-14ths IN THE 4-3-4-3 'X' SHAPE AND THE 5-4-5 TRINE, DOES NOT HAPPEN EVERY GALACTIC ALIGNMENT, THE STOP AND START POINT OF THE GREAT YEAR…!!!

During the string of Great Years, that come and go through the vast expanse of time, the sacred geometry of the 4-3-4-3 'X' and the 5-4-5 TRINE, is not always occurring at the same times. For that to happen the Celestial Axis and its Equator, as well as the Ecliptic Axis and its Equator, would need to remain in their positions constantly, and they don't. These two, the Celestial and the Ecliptic, move just slightly over long periods of time. And even though it is just slightly, it is enough to change that shape from its true 4-3-4-3 'X' shape. There are also several factors involved about stellar events in our solar system besides the 25,920 years in the Great Year of Precession. And they too control when the Galactic and Ecliptic Axis make the 4-3-4-3 'X' aligned with the Solstices. Altogether, it shows that the here and now of our current Era is a perfect alignment for Sacred Geometry in the 2nd and 3rd dimension. And this Conjunction of Spheres probably happens only once in infinity.

Chapter 36: The Full Celestial Cycle

The Celestial Plane and Axis do not stay in place but move back and forth in a cycle. The Celestial Axis and its Equator move from 22.1 degrees to 24.5 degrees away from the Ecliptic and then back again to 22.1 degrees.

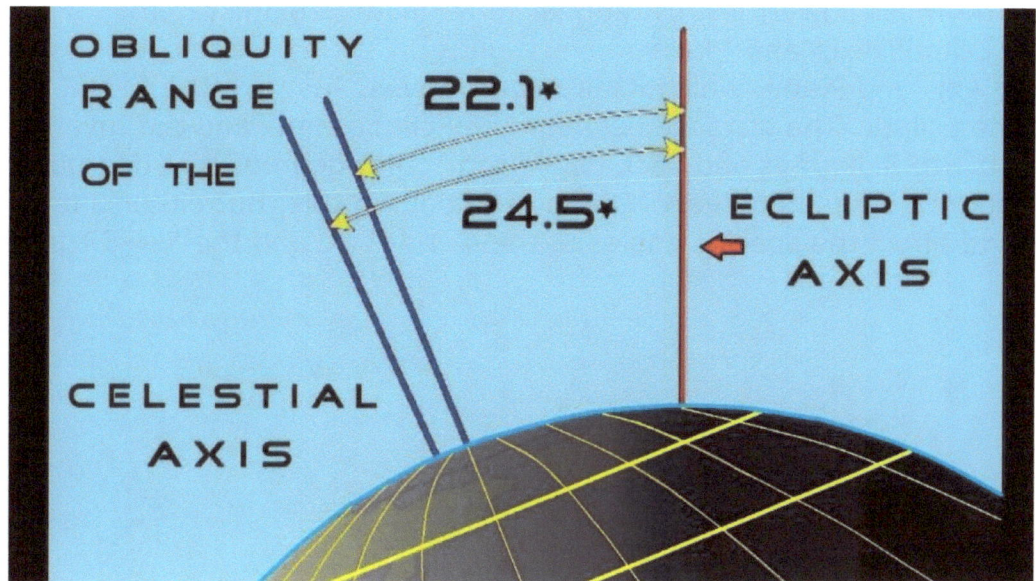

This takes a 41,000-year cycle and that's a full 2.4 degrees. Currently, the Earth's obliquity is 23.44 degrees, and this is very close to 23.3 degrees, which is the center mark between the two extremes.

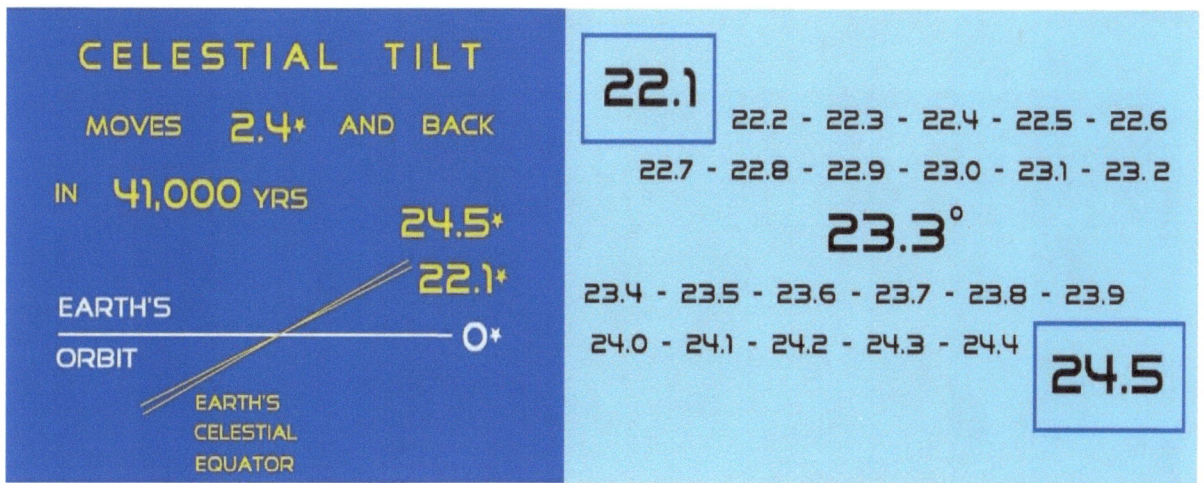

Chapter 37: The Full Ecliptic Tilt Cycle

The second tiny movement is how the Ecliptic Axis and Plane move slightly. The Ecliptic Plane tilts a tiny amount in a 70,000-year cycle, just like the Celestial Plane in its 41,000 years. And as the Celestial Equator is measured off from the Ecliptic Equator; the Ecliptic Equator is measured off from Jupiter's Invariable Plane; which is the most powerful gravitational force in the solar system, after the sun. Also, the Ecliptic Axis and Plane move slightly in a tiny wobble through Jupiter's Invariable Axis and Plane, which itself never moves. Just like the Celestial wobbles through the Ecliptic. The distance between the Ecliptic Plane and the Invariable Plane is around 0.1 to 3 degrees and back (vcalc.com), or 0 degrees to 4 degrees and back again (boek.nl). I could find very little information on this. But either 3 or 4 degrees is enough to affect the specific angles of the 4-3-4-3 'X' and the 5-4-5 trine.

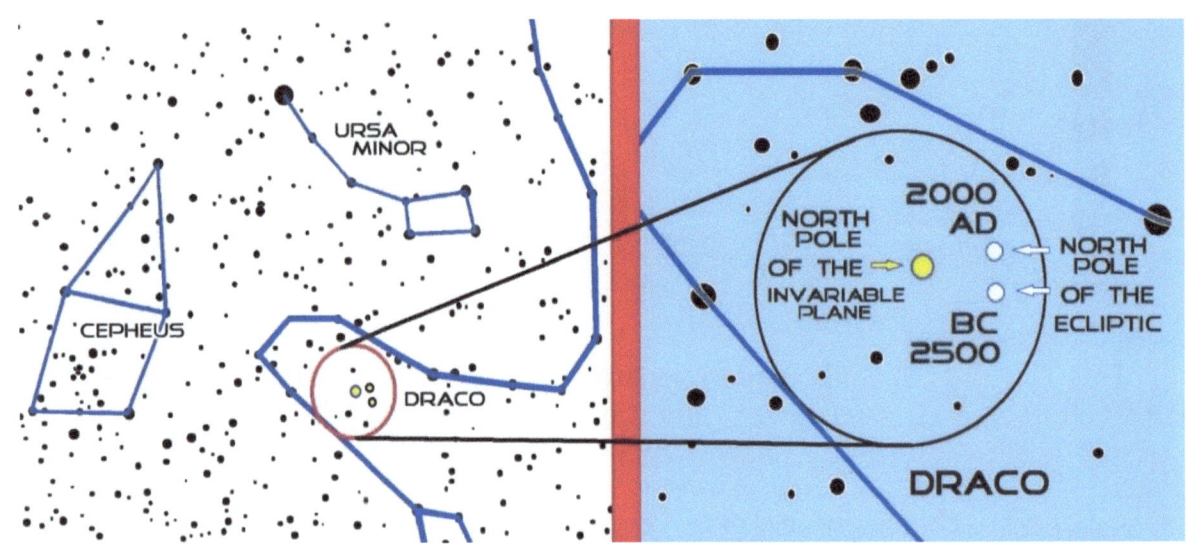

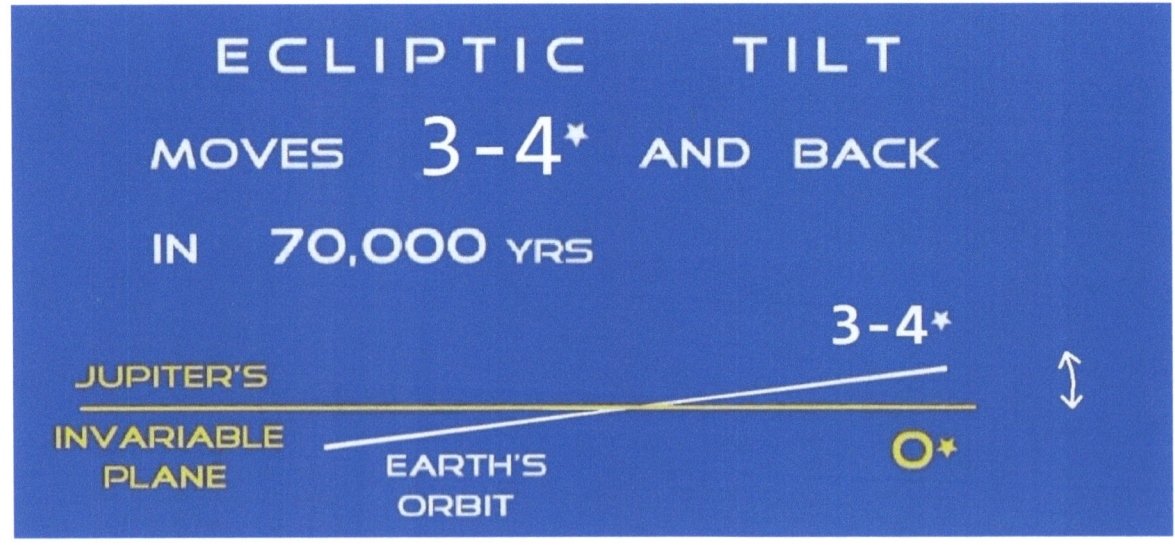

UNDERSTANDING REVELATIONS BY ASTRONOMY

Chapter 38: January 1st & July 3rd

The next alignment is how there are several groups of cycles that occur right now on or close to January 1st and July 3rd. The gravitational forces collected together, in alignment; at those times right now; near to, or on those dates; is to me, simply astounding.

The Earth's orbit has its Solstices and Equinoxes that occur, more or less, on the same dates every solar year. During Precession those dates will move through different stars and stellar features that are not tied to the solar orbit. This is why the stellar feature of the Galactic Center is now in conjunction with the December Solstice sun. When the sun crosses the sky on that day, the Galactic Center is precisely behind it. 72 years before now, this alignment happened on December 22; and another 72 years earlier the alignment happened on December 23. And so on, and so on, this is Precession.

But when discussing the solar orbit of Solstices and Equinoxes; that moves through the starry sphere, to include the Galactic Center; there is another very important stellar feature which moves through the solar year dates. That's when the moving Celestial Equator twice crosses the unmoving Galactic Equator. These two nexus points happen right now in Precession, I believe, on the dates of January 1st and July 3rd. And it is in a 4-3-4-3 'X' alignment with the Equinoxes.

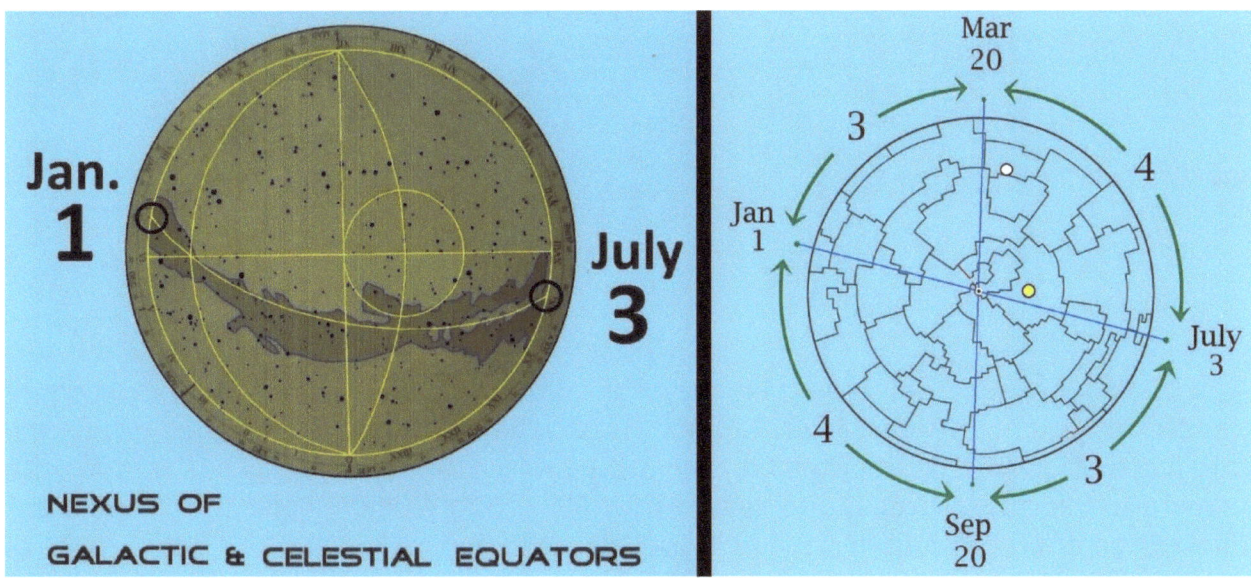

UNDERSTANDING REVELATIONS BY ASTRONOMY

Chapter 39: Jupiter & it's Invariable Plane

In our solar system, the planet Jupiter weighs more than all the other planets combined. And Jupiter, with its moons; is not alone in its 12 year orbit that takes it around the sun. Jupiter has two collections of planetoids in its orbit path. These three all move at the same speed in their crowded orbit, so they never collide with each other. But this means that there is more weight than just Jupiter in its orbital plane. A lot more weight than just Jupiter.

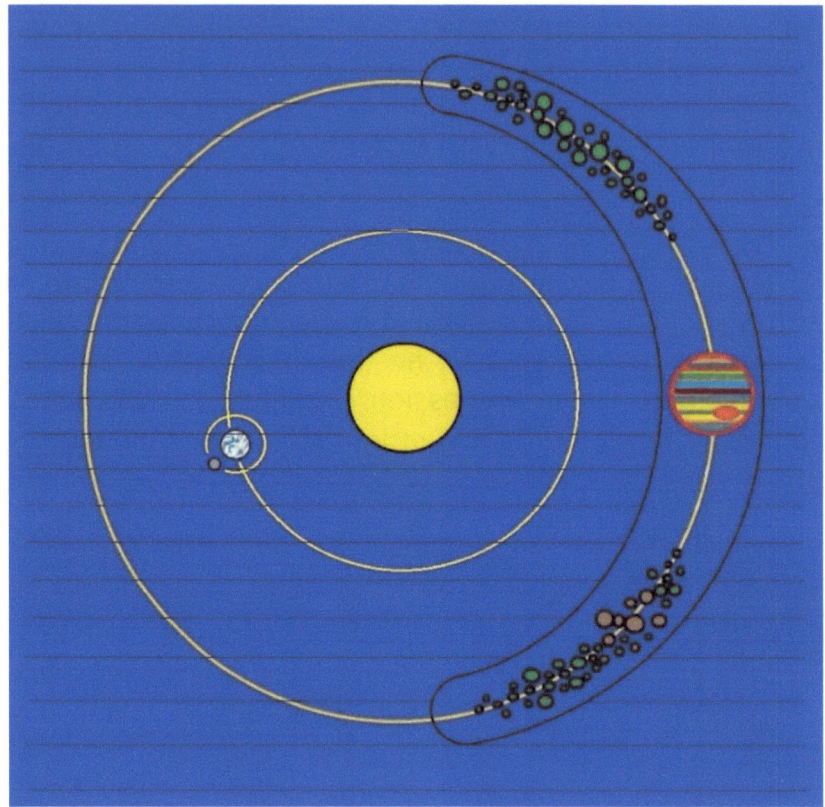

All this gravitational force on the orbit plane of Jupiter and its two clusters of planetoids, creates what's called, Jupiter's Invariable Plane. It never varies or moves. This force of gravity is so powerful that it pulls the other planet's orbits in line with its own, to varying degrees, and even more. All throughout the solar system there is a collection of stellar rubble created by the plane of Jupiter's orbit. It's a thin layer of relatively small rocks, lots of rocks, filling the entire solar system like a sheet of paper.

This Invariable Axis and Plane does not have a slight titling back and forth, like the Earth's Celestial and Ecliptic Axis and Planes do; once again, it never moves. That's, why it's called Jupiter's Invariable Plane. And its force is the most powerful affect in the entire solar system. Remember that even without the two enormous clusters of planetoids, Jupiter alone weighs more than all the other planets combined.

The Earth's orbit (the Ecliptic Plane) is slightly tilted away from the Invariable Plane by 1.57 degrees. Every 100,000 years the Ecliptic Plane makes a wobbling spin around the Invariable Plane. This means that there are always 2 dates, roughly six months apart, where the plane of the Ecliptic passes through the Jupiter's Invariable plane. Right now the Earth's Ecliptic orbit passes through Jupiter's Invariable plane on January 9th and July 9th.

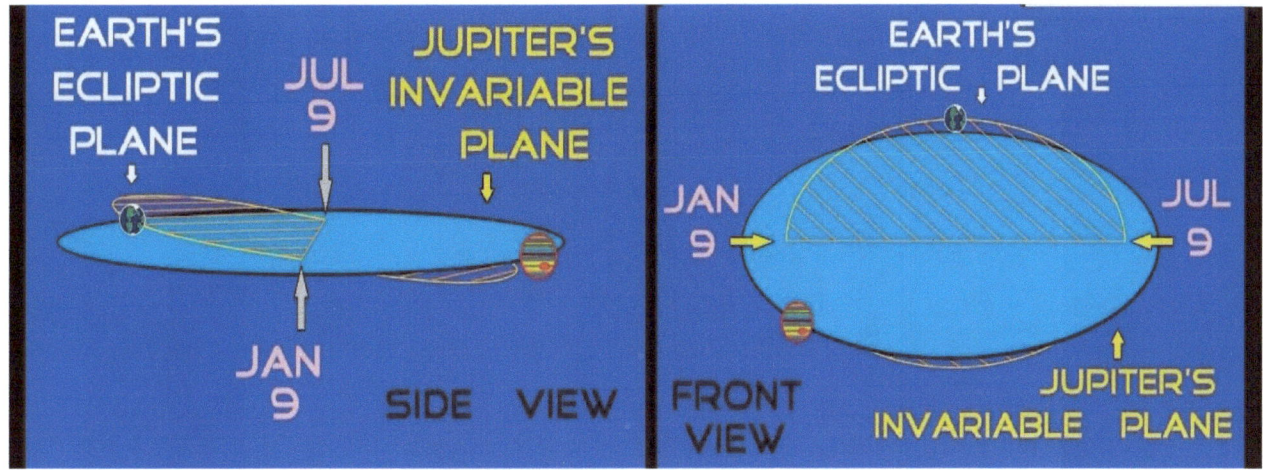

Because of that we have meteor showers on those dates; from that thin sheet of rocks, the stellar rubble. As the Earth's Ecliptic orbit revolves, the slight variation of its own tilting degree measurement, has a 70,000-year cycle. But also; that whole plane revolves around through Invariable Plane in a 100,000-year cycle. Which means that in another 100,000 years the Earth's Ecliptic Plane will pass through Jupiter's Invariable Plane on these same dates. And as the Celestial precesses one degree in 72 years (currently 71.6 years), through the Ecliptic; the Ecliptic itself precesses one degree in 277.77 years, through Jupiter's Invariable Plane (approximately 274 years for every day).

....

Chapter 40: The Shape of the Earth's Orbit

The Earth's orbit has -1) a 21,000-year cycle, and -2) another 100,000-year cycle. For the first one, the sun's position is never in the exact center of the Earth's orbit. In fact, it's a slightly egg-shaped orbit; so there will always be two dates, that the sun and Earth are closest (the Perihelion), and farthest (the Apthelion); from each other. These dates move through the calendar one day in 58 years (timeanddate.com).

Right now, those two rotating dates, when the Earth is farthest and closest to the sun; happen around January 1st to January 4th, and July 2nd to July 5th. This egg-shaped orbit revolves completely around in roughly 21,000 years. It combines a 19,000-year and a 23,000-year cycle (Dr. Krupp).

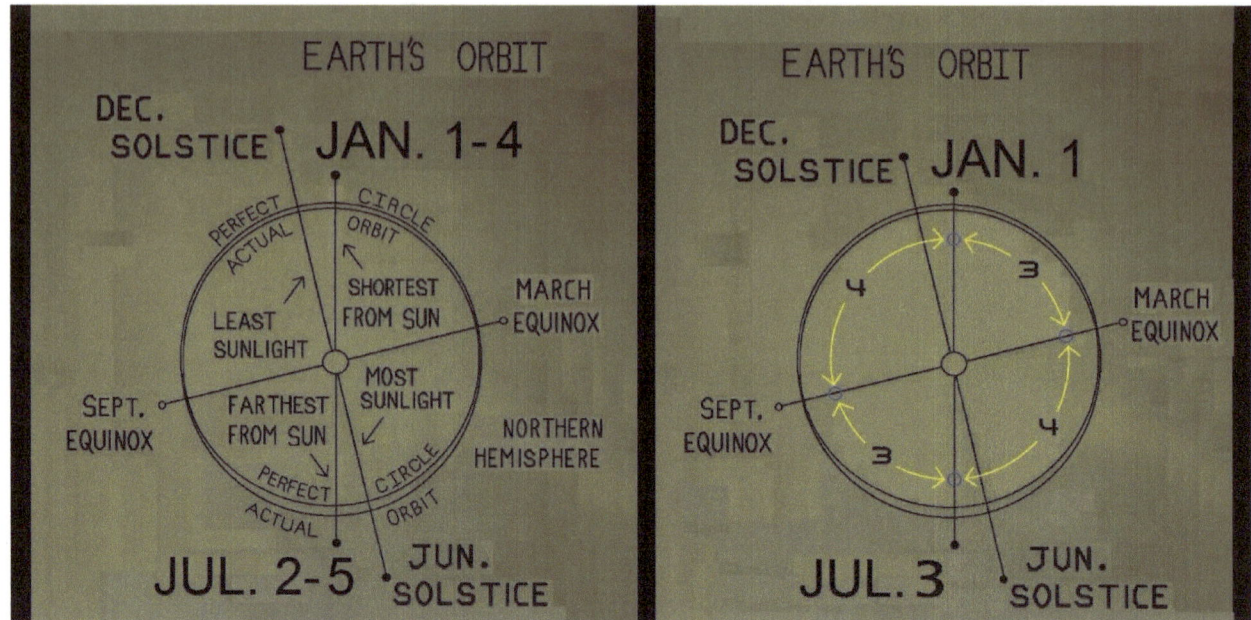

Those two solar dates, from when the Earth is farthest and closest to the sun revolve around every day of the solar calendar. In our Stop/Start Era now, those dates are around Jan. 1-4 and July 2-5; roughly six months apart. And it is the Jan 1st and July 3rd that holds the 4-3-4-3 'X' with the Equinoxes. In the year 1246, the earth was farthest from the sun, the Perihelion, on the December Solstice. And it is estimated that in 6430 AD it will be on the March Equinox (timeanddate.com). These hellion dates move the opposite direction of Precession, which is the same direction as the solar orbit of the calendar year (Dr. Krupp).

And also, not just the direction of the egg-shape of the planet's orbit changes. In 100,000 years the shape of the Earth's orbit changes. It goes from a near circle, all the way to a stretched out oval shape, and back again to a near circle. It is called the 'Eccentricity' of the Earth's solar orbit.

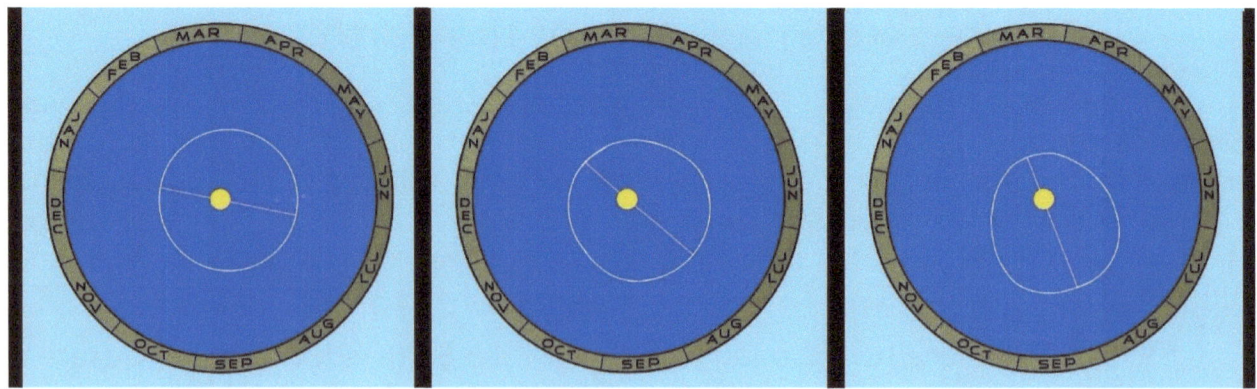

UNDERSTANDING REVELATIONS BY ASTRONOMY

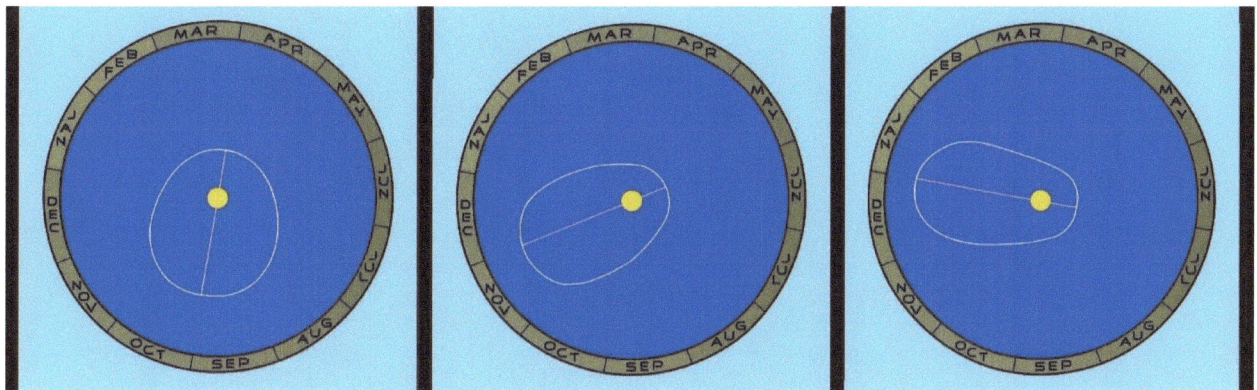

We are very close, right now, to the circular shape. (My diagram is exaggerated to show the long oval.) But this revolving 100,000-year egg-shape of the Earth's orbit does not change how the wobbling Celestial Equator precesses through it 25,920 years; they are two separate events.

....

Chapter 41: And Now All of it Together

Astronomically, there are many cycles that create the full Galactic Alignment going on right now, and it is complicated. It is more than just the 25,920 years in the Great Year of the Precession of the Solstices and Equinoxes.

In the field of sacred geometry all the ancient cultures of the Old World believed that the formulas for the Cubed-Sphere and the Circled-Square were the most perfect expression of divine orchestration for both two dimensional and three dimensional realities. And that's what's so special about this Great Year out of all the other cycles of Precession.

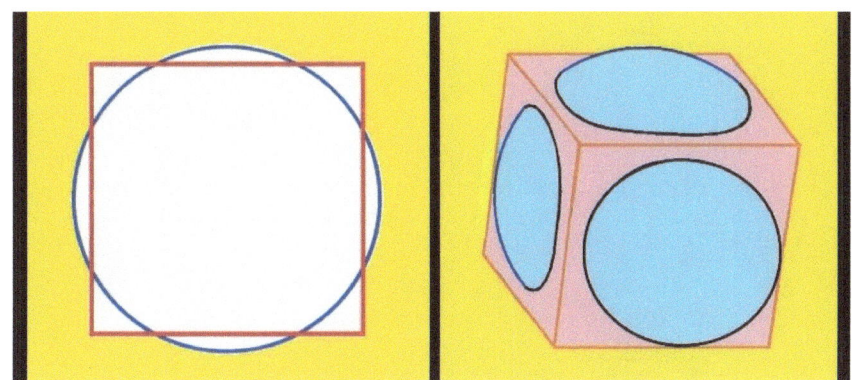

And the measurements of $1/7^{th} - 2/14^{ths}$ and $2/7^{ths} - 4/14^{ths}$, are used to create the volume of the Cubed-Sphere and the perimeter of the Circled-Square; as is shown on the Masonic Cross diagram.

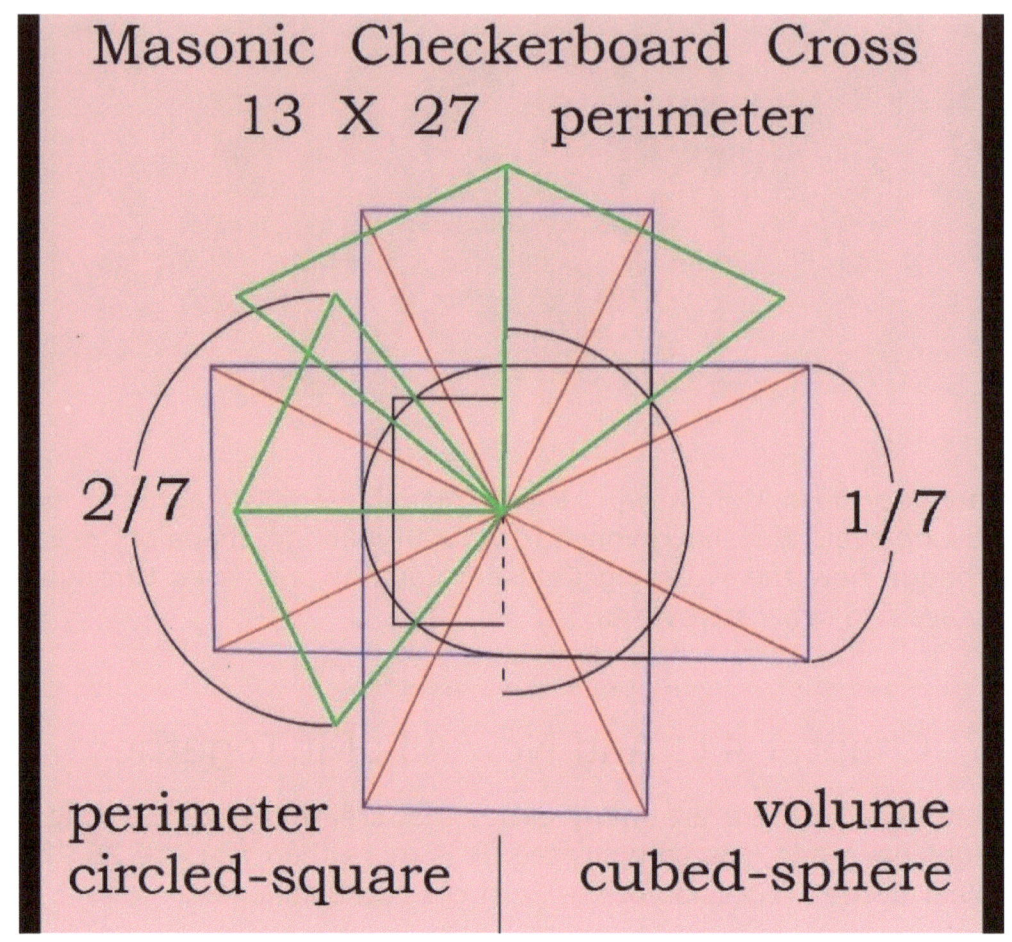

And specifically, the 4/14ths – 2/7ths measure is created from a 14-fold division of the circle two ways. First, there is a 4-3-4-3 'X' shape; and second, there is a 5-4-5 trine pattern, like a triangle.

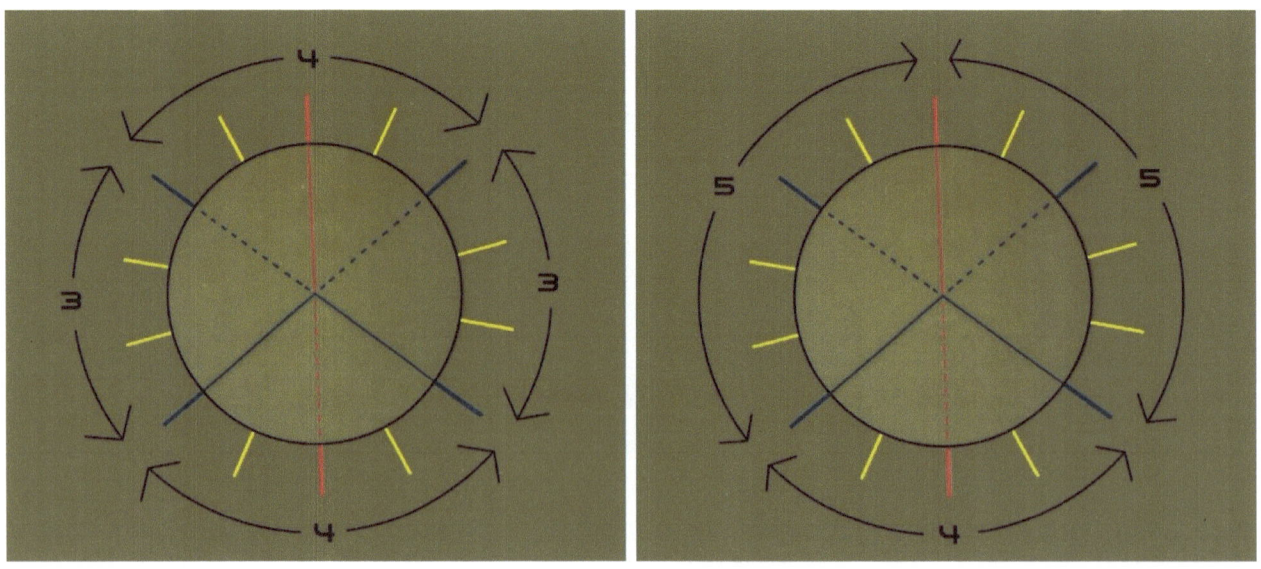

UNDERSTANDING REVELATIONS BY ASTRONOMY

This was encoded in many of the worlds mythos, especially the *Revelations* of Saint John the Divine. With the Dragon and Beast's 14 heads, the heads without horns is 2/7ths and the heads without crowns is 1/7th.

Right now, in the Great Year, at the stop-start for the 25,920 years of Precession; the view sighted straight down the Celestial Axis creates the illusion of the Galactic Axis rotating around the Ecliptic Axis, which makes the 4-3-4-3 'X' shape.

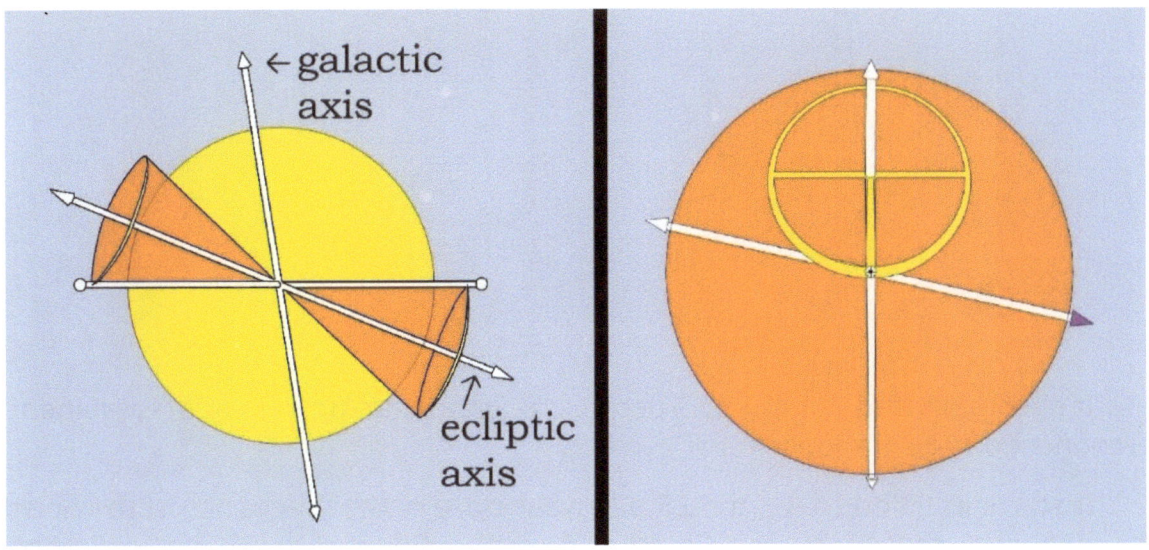

Both the 4-3-4-3 'X' shape and the 5-4-5 trine pattern have several synchronizations to key dates in the calendar. First, with the Galactic Axis and Ecliptic Axis. As well as, the 260-day Sacred Calendar inside the Solar Orbit.

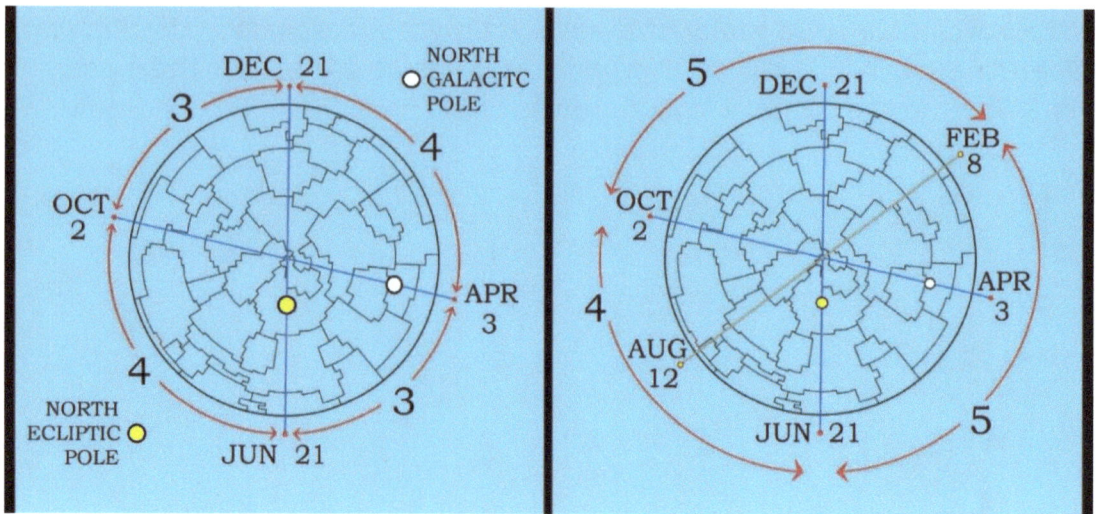

The Equinoxes dates have a 4-3-4-3 'X' alignment with the Jan. 1st and July 3rd crossing of the Galactic and Celestial Equators, when they are due north/south at noon or midnight, called the Celestial Meridian. And there are other events centered around Jan. 1st and July 3rd.

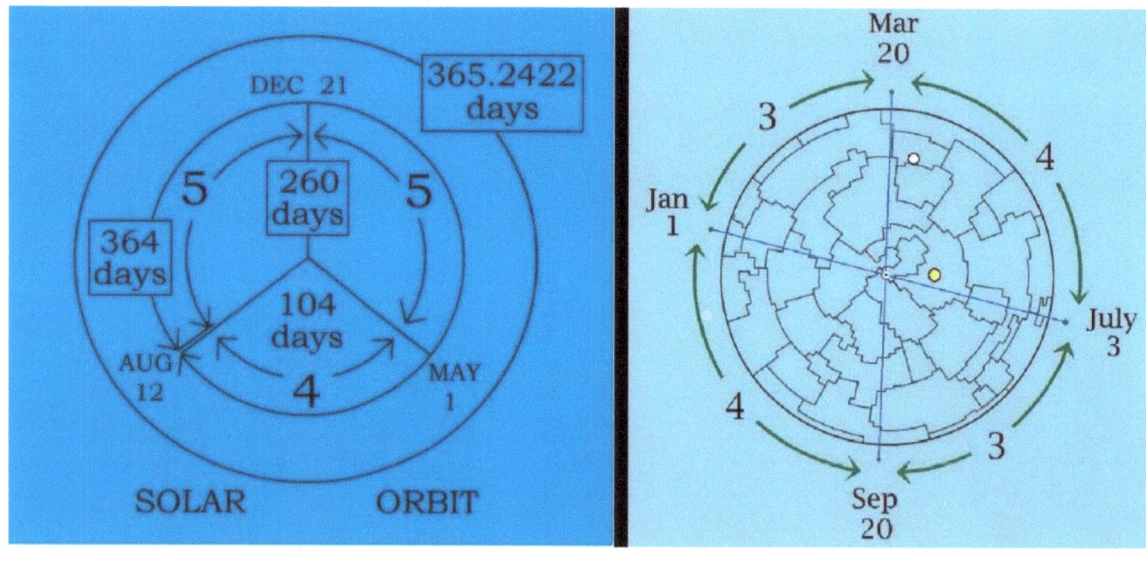

....

It's complicated for sure. There is a lot going on. I have explained the interconnection of 7 various stellar cycles.

It starts as follows: -1), the 25,920-year cycle of the Precession of the Solstices and Equinoxes, caused by the Celestial Axis and Equator swiveling in Precession around the Ecliptic Axis and Equator. -2), There is a 41,000-year cycle of the 2.4 degree tilting variance for the Celestial Axis and Equator. -3), And there is a 70,000-year cycle for the 1.8 degree tilting variance of the Ecliptic Axis and Equator. The 41 and 70 thousand year cycles do not have dates that rotate through the calendar year.

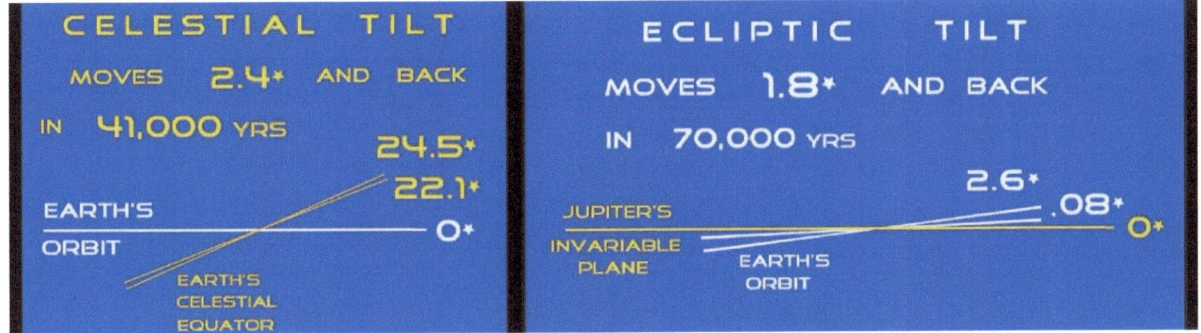

There are also the stellar events near or on January 1st and July 3rd; -4), There is the 25,920-year cycle for the nexus of the Galactic and Celestial Equators on the Celestial Meridian. -5), There is the 100,000-year cycle as the Ecliptic Plane swivels through Jupiter's stable Invariable Plane, now on January 9th and July 9th.

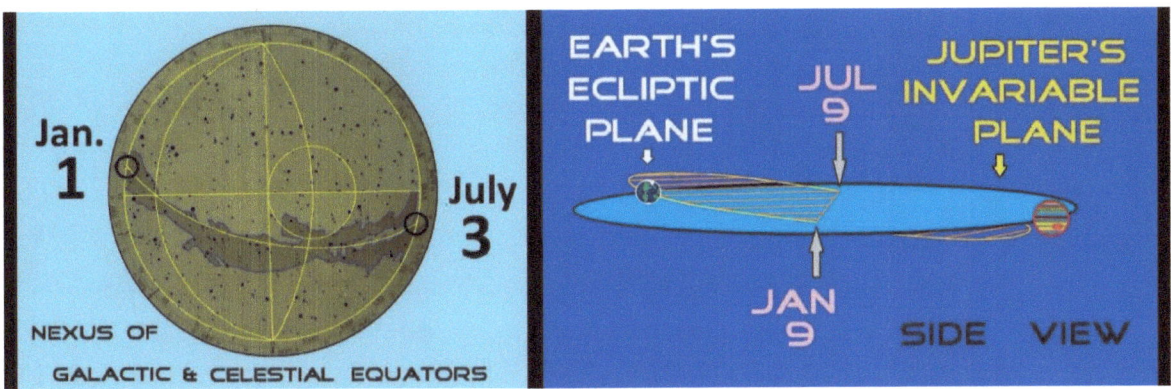

-6), There is another 100,000-year cycle of the rotating dates. The Earth's orbit goes from a near circle stretched to an oval; and back.

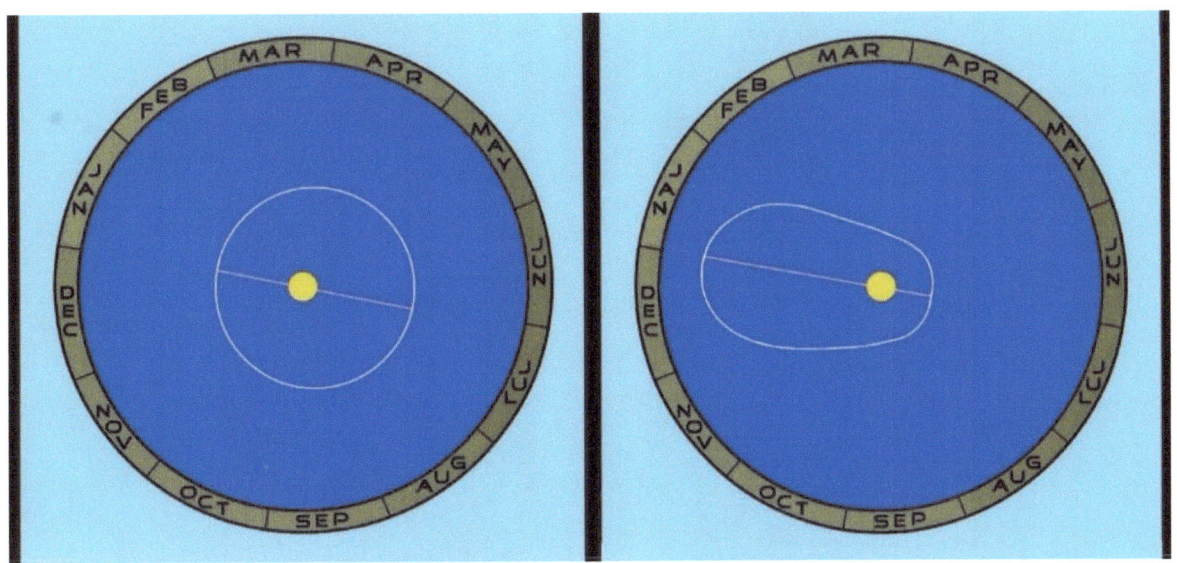

-7) There is a 21,000-year cycle of the rotating dates when the Earth is closest and farthest from the sun. And currently these dates are on or near the July 3rd and January 1st. On a hemispherical star chart, the dates of January 1st and July 3rd create the 4-3-4-3 'X' with the March and September Equinoxes.

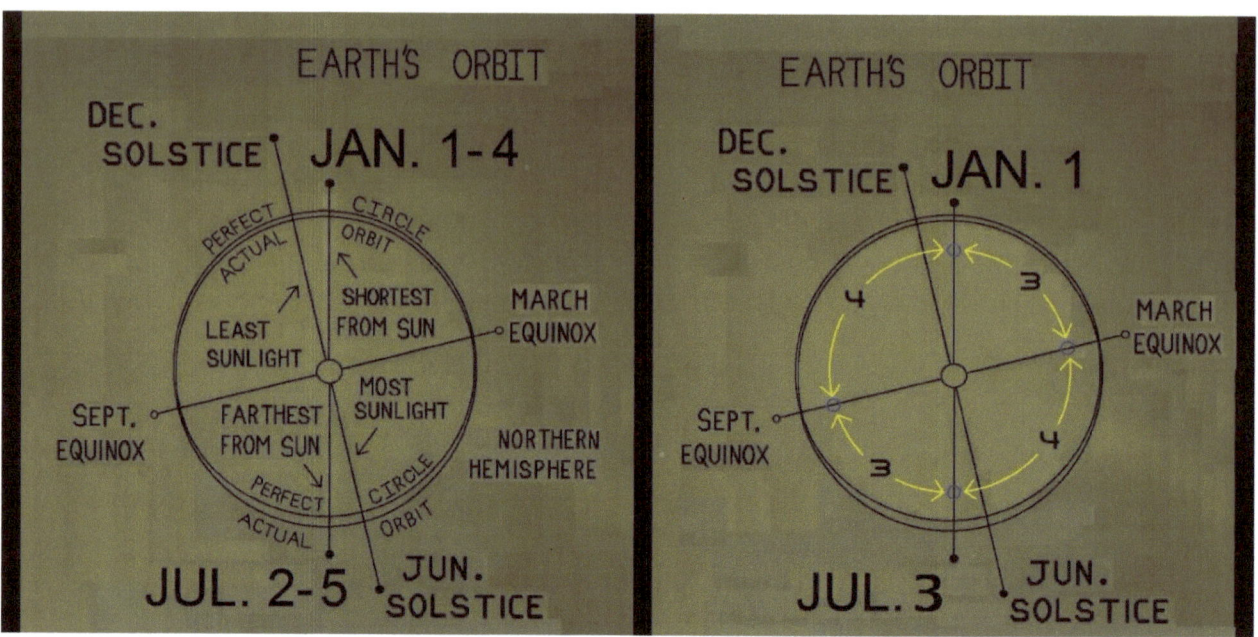

As I said earlier, to me, the gravitational forces collected on and around these two dates of January 1st and July 3rd are simply astounding.

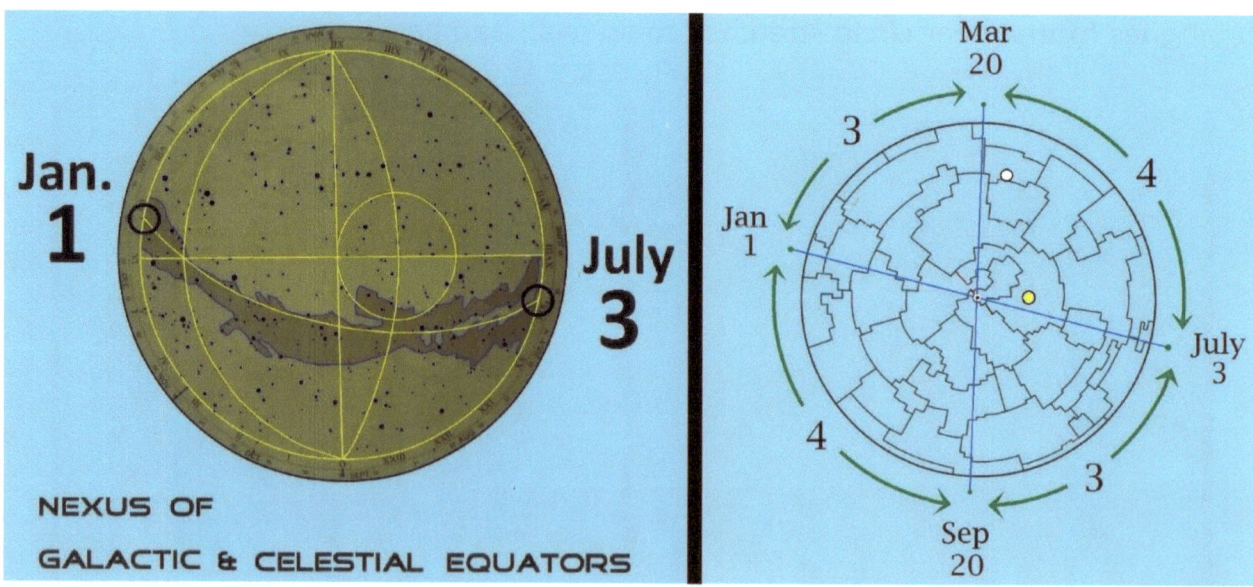

Precession moves the twin dates when the nexus of the Galactic and the Celestial Equator are on the Meridian. Right now in our era they align with the Meridian on Jan. 1 and July 3. There are more cycles besides Precession that center on these dates. But also, both the 70,000-year Ecliptic Variance cycle, and the 41,000- year Celestial Variance cycle; are not fixed to calendar dates. Below is a diagram of all the other stellar events centered around January 1st and July 3rd. It shows the difference between the twin dates, 6 months apart, for how close the alignments are. If the favored day of the Messiah's birth is July 3 over January 1, then perhaps it was thought so by the ancients because Jan. 1 holds the daylight, and July 3 shows the constellations at night. Perhaps it was ordained by the divine because July 3 has a tighter group of alignments than its twin date.

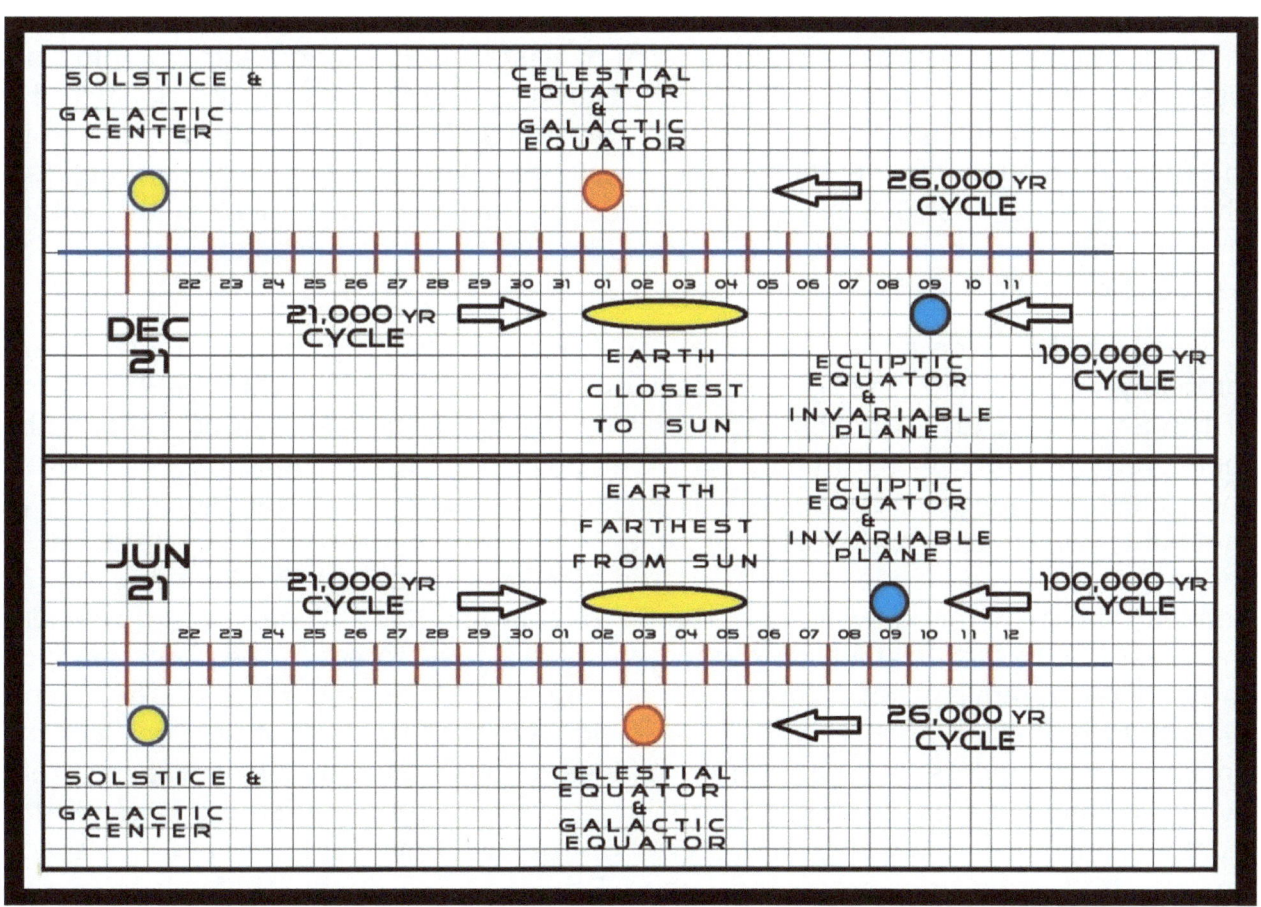

The cycles of time are this... The one stop-start point for the 26,000-year cycle, for Precession & the Meridian nexus of the Celestial and Galactic Equators... -X- the 41,000 years for the Celestial Variance... -X- the 70,000 years for the Ecliptic Variance... -X- the 100,000 years for the orbit rotating through the Invariable plane & the orbit shape... -X- the 21,000 years for the dates closest and farthest from the sun = 156,702,000 years.

There is the illusion of the 2/7ths - 4/14ths divide between the Ecliptic and Galactic Axis, sighted down the Celestial Axis. And with the Masonic Cross, this creates the measures of the interior volume of the Cubed-Sphere and the perimeter of the Circled-Square.

Perhaps this complex event happens again sometime before all the cycles of the event come to coincidence again, perhaps not. Regardless, I believe we are talking about an alignment of cycles much greater than 25,920 years. How much so? I would love to run very specific variables of these cycles into a supercomputer. It would be fascinating to find out, if and when, there could be another Sacred Geometry Alignment of the Cubed-Sphere and the Circled-Square like this; either before or again after now.

Is this why Julius Caesar Changed the New Year day from the Equinox on March 21st to Jan 1st? I believe so. In ancient Alexandria there was the Egyptian based mystery religion of Serapsis, it was a form of Isis/Ast and Osiris/Ashir worship. Isis was said to have delivered her first child, Horus on the December 21st Solstice. He was the child for the solar year. And then on the twelfth day after she gave birth to the Baby Aion who was the child of the Great Year. And that would be January 1st. There is a belief that Julius Caesar learned his calendar system from the Mithra Mystery cult. And their famous art work is known to have constellation animals at key positions.

Most likely, the advanced cycles of 21, 41, 70, and 100 thousand years were 'probably' not known by the ancients. But, it is certainly believable that the Precession of the Solstices and Equinoxes had several ways for its almost 26,000 years to be measured. And one of them could very well have been the rotating dates for when the Celestial Equator crossing with the Galactic Equator was due north/south at midnight; which is the Celestial Meridian. Is this why the U.S. Declaration of Independence is bookended with the dates of July 2nd and July 4th? To mark July 3rd between them? Is the knowledge of the 5-4-5 split and the 260-day sacred calendar the reason why the Washington monument is designed to catch the sun's passage through its pyramidon on August 10th or August 11th, as reported in *The Secret Architecture of Our Nation's Capital* (by David Ovason)? I believe it was not just the ancients who understood parts of this Precessional Cosmology. There is great evidence that some of the United States of America's Founding Fathers knew of this from Masonic lore. And while I discovered on my own how the Masonic Cross shape creates the Circled-Square and the Cubed-Sphere, it is probable that this is part of the Masonic Lost Lore. And it seems certain it was encoded into *Revelations* by Saint John the Divine Revelator, and *Genesis* itself.

PART IV – The Art of the Fugue

Let's talk about what's been hiding in the human mind! And hiding in plain sight.

How do you leave a message for the future, so it lasts after you have been dead for hundreds of years, or even thousands? You make a story that will get told over and over as the generations pass; and it will not get changed in all that time. And that's the tricky part, how to keep your message from getting changed. Even sacred texts get changed, at least the Bible did, over and over, and over again. The tricky answer is… You leave a story that people don't know is a story. Welcome to the 'Art of the Fugue'.

....

Chapter 42: The Dictionary Definition

The first way to explain the Fugue is with the 2 dictionary definitions.

....

From the movie, *Pirates of the Caribbean, Curse of the Black Pearl*… In the pirate's watery cave, on the 'Dreaded' Isla de Muerta; Capt. Jack wakes up from after Wil Turner knocks him out with the steering oar. So, Jack is in a daze for a few minutes. As he gets caught by Barbosa's crew, Capt. Jack gets stuck in a fugue trying to remember the word 'parley'. The ship's crew disbelieve he's still alive and tell him he should be dead.

Capt. Jack – "Am I not?... Parliley, parli, parlilue, parsnip, parsley, par, partner, partner…"

Ragetti – "Parley?"

Capt. Jack – "That's the one parley, parley!"

This is the 1st dictionary definition of Fugue; A state of confusion where things are not remembered, or things are not seen as they really are. As a motif in the Art of the Fugue, it's crazy-sly like Hamlet, or crazy-delusional like Don Quixote. Capt. Jack's Fugue confusion is placed in the overt text for the -passing of words, the parley. It's a partnership between the overt text and the secret encoded text. It's a partnership of words, an understanding.

....

Capt. Jack sits on the deserted island at night with Miss Swan drinking Rum. She makes a toast to the Black Pearl. The very drunken Jack tells her the soul of the ship to be able to go wherever they want to go.

Capt. Jack – "…That's what a ship is you know. It's more than a keel, and a hull, and a deck, and sails. That's what a ship needs. But what a ship is, but what the Black Pearl really is… is freedom."

The 2nd dictionary definition of Fugue is the musical reference. A Fugue composition is where each separate part of the whole takes the lead for a short time in its place. This is also an homage to *Hamlets Mill*, (p.49, HM) with its similar commentary on "The Ship that Found Herself", by Rudyard Kipling. It's about how the parts of the ship, are not the sum of the ship.

It's motif is to represent all the various topics and parts to the Art of the Fugue. The metaphor turns all the various music pieces of the Fugue composition, into jigsaw puzzle pieces. In another description, we have the many strands of the spider web. This is the 'Big Picture' topic of the Fugue, that ties together all the other various topics.

Chapter 43: *Hamlet's Mill*

The Art of the Fugue is a specific language of metaphors and motifs; THAT IS HIDDEN BETWEEN THE LINES... so it is kept secret and it gets preserved through the ages. But... who preserved it? And for how long? We don't know who the first encoders where because it was just too far back. But it was definitely an intellectual, stargazing elite; possibly the priesthood, or an organization of traveling traders, either by land or sea. It was whoever made so many of the world's most ancient mythologies and fairy tales, including the Sumerians.

This method of transmitting encoded information into oral traditions may go back to the Late Neolithic Age, the so-called -Revolution, around 5000-6000 BC. So, this Art of the Fugue is very ancient and world-wide. But it must be stressed; that however it happened; there is a continuity of the same motifs for all these different, ancient, world mythos that are spread through thousands of years. And for truth, who knows when the ancient mythos were first spoken as oral tradition?

The first book I know of to OPENLY talk about this in modern times is called, *Hamlet's Mill*, published first in 1968. And it weaves a tale of what the Art of the Fugue is by explaining what it encodes and how it does it in the most ancient worldwide mythos. The authors of *Hamlet's Mill*, Hertha von Dechend and Giorgio de Santillana, tell us that the ancient Greeks 'rediscovered' this Art at the beginning of their Golden Age. By the time of Plato, it had been studied for over 400 years. "Plato did not invent his myths, he used them in the right context -now and then mockingly- without divulging their precise meaning: would tell cryptic tales, whoever was entitled to the knowledge of the proper terminology would understand them." (p.310 HM)

Chapter 44: The Fugue is Older than Dirt

The oldest reference to 7 and 7, that I know of, goes back to the Akkadian and Sumerian mythos. The axe of Gilgamesh weighs 7 talents and 7 minas in the mythos called, *The Huluppu Tree*. And in *Inanna's Decent into the Underworld*, she leaves 7 items at 7 gates; as translated in *Inanna, Queen of Heaven and Earth*, (by Diane Wolkstein).

These mythos are at least from 5 to 6 THOUSAND years old; holding a secret for HERE AND NOW IN TIME! They are about the Galactic Alignment that is happening right now! They might even be older; and could have been preserved in oral tradition for thousands of years before they were pressed into the clay tablet by the reed stylus.

Culture and Mythos passed from Sumer to Arcadia. One of the oldest stories we have, the *Enuma Elish,* (trans. by Leonard W. King); favors Marduk, the son of Ea/Enki, as the head god. It is Marduk who wins the battle, gets 'raised- up', and is given titles of power by the other gods. But the earliest tablets say that it was Ea/Enki, not Marduk, as the triumphant god raised up in victory. Many stories stayed the same, replacing one god for another. Also, stories where changed, by a new people adopting an older culture's beliefs. Stories also change in the same culture over time. Sumerians were the first to write about Adam and Eve; but it is very different from the modern Bible.

....

Chapter 45: The Visual Story

And all of this was planned, ages and ages ago, by some world-wide cultural influence that encoded specific topics disguised as metaphors and actions, and visual symbols. In *Don Quixote,* (by Miguel de Cervantes) The hero tilts his lance at windmills. The lance here is the Ecliptic Axis. The windmill with its 4 spinning blades, represents the solar orbit with the 2 Solstices and 2 Equinoxes. It has 2 meanings in astronomy; both of which are tied to the Ecliptic Plane. The 4 spinning blades have 2 different levels of understanding. First it is for the solar year, and the fan blades represent the calendar year tied with the 2 Solstices and 2 Equinoxes. And so the calendar year turns.

But the 'Quixote delusion' is that the earth stays in one place and it's the calendar with its 2 Solstices and 2 Equinoxes that move. When in reality, it's not the calendar that moves, it's the earth moving through its orbit. For the second meaning, it's still the 4-armed cross of the Solstices and Equinoxes as it spins backwards, ever-so-very- slowly, with the 25,920 years of the Precession. Hollywood loves the slowly spinning fan blades, *Galaxy Quest* used the full motif best.

This visual motif is just one of the many symbols and physical actions, presented by *Hamlet's Mill*. Certainly, the lance titling at the windmill is a great visual image together with the 'action' of the spinning blades. Sometimes, the clues are visual images of objects that people can easily imagine; like the jaw-bone of the long snouted beast. The jawbone has no action, it's just an image of part of the visible Milky Way. Other times its less of an image and it's more of a specific action like a 'Fall' straight down, the longer the fall the better. Obviously, the straight line of the drop is determined by the gravity radiating out from the center of the planet. And this action does very well to portray the symbol of an 'axis' through the force of gravity.

And most important, the same symbol can have different meanings for different topics. So, another meaning for 'the Fall' is the fall into death. Just remember that the Art of the Fugue has several different sets of topics. And the Fugue's topics, of encoded motifs, extends beyond the knowledge of the Precession of Solstices and Equinoxes.

Chapter 46: The Jam of Preserves

Secrets were encoded by a very extended group of mysterious 'over-see-ers', of human culture. They had advanced astronomical and calendric information about the timing, for the then, upcoming Galactic Alignment. This knowledge was encoded into mythos as, the Messiah at the End of Times; here and now at the stop/start for the 'Great Year' of Precession. And I hope by now I have adequately explained how things were encoded; and what is Precessional Cosmology; also how these sets of topics make the 'big picture'.

This Art of the Fugue was preserved in cultural treasures through the ages, in stories about cultural gods and heroes, and stories about the constellations. All so they would never get lost. Thus, when the truth came out, there would be no doubt that the entire ancient world was in on the secret. It's in the Greek epic of *The Odyssey*. The book *Homer's Secret Iliad* (Wood & Wood), gives evidence that it's in the Trojan War epic itself; the entire first part of the *Iliad*, is a list of ships and sailors; which is a catalogue for specific constellations and all their stars. The big stars were 'named' heroes, and for the tiny stars seen by the naked eye, these 'troops' were counted as a number.

The Art of the Fugue must have covered topics taught in the many Mystery Religions of the ancient world. And I believe that by the time of John the Baptist, Jesus, and Magdalene; many people all over the ancient world must have known parts of this secret knowledge. Many people knew, either a little, or knew a lot, about the Fugue. And then that changed. I think it was because, if too many people KNOW ABOUT IT; then to many people can tamper with its message. So, the secret was made to be forgotten.

Chapter 47: The Fugue Gets Lost & Found

For whatever reason, it seems that the Art of the Fugue was erased from public knowledge... And now here in the book, it's easy to point fingers at the Church of Rome. It was they who erased and snuffed out all the overt knowledge of the Fugue and its topics. So for 1000 years all knowledge of the Fugue was being 'lost', at least in Europe and the Mediterranean.

By 1100, the Templars grew powerful in Europe. They are the Great Grandfather of all the Western mystery traditions -900 years ago, to present. It was they who rediscovered the 'Language of the Birds', this Art of the Fugue.

The Fugue re-awoke and started its magnificent flourishing in the Romance Era of literature. It's quite possible the Islamic world may have remembered the Fugue and passed it on to the Templars.

And of course, those Templars had a secret mission in the Holy Land. According to Masonic lore, and also in Dan Brown's, *The DaVinci Code*; those original 7, then 9 knights, dug up a lot of treasure underneath King Solomon's temple. Reportedly, they also dug up preserved scrolls of texts and really old maps; lots and lots of them. Enough scrolls to fill four chests, with each chest so large it took six grown men to carry one. This underground library was put there supposedly by the learned Jewish mystery sect, the Essenes. They secretly did this when they did the deconstruction and rebuilding of Solomon's Temple, for Herod the Great.

This library was filled with scholarly knowledge. Such as advanced architecture; and other sciences, like medical knowledge. And it was filled with religious texts, of which so many are now lost. Texts from; the Greek Orphites, and the Jewish Gnostics; of which Christian Gnostics grew out of. According to Masonic lore, there were also scrolls of some early Johanite gospels; from those who believed that John the Baptist was a Messiah more powerful than Jesus. The Templars also would have found some of the older Jewish texts like; the Book of Jasher, the Book of Enoch, and the Book of Adam and Eve.

The Templars would have found Precessional Cosmology; because a wandering dessert people know the stars. And they also would have found, the Art of the Fugue. They would have found explanations and examples of all the topics. They would have read all about the actions and images. Here is a modern example of how the same types of actions can be encoded into any story at all... First, Capt. Jack Sparrow jumps off the top of the 'mast/pillar', the axis. Then he 'falls' down into the ship taking command. Last, he starts bailing water out, as the Flood Motif. So, that's the clues. The event, is that the Old World Age has ended and a New World Age begins.

Here is the story, with the Fugue language interpretation as per *Hamlet's Mill*. It starts as a young upstart becomes the New Celestial God. He does this by momentarily taking the place of the older, more powerful Ecliptic God. So, the hero climbs the Ecliptic Axis, with his Fall. And then, from the New God's position on the Ecliptic Axis he saw everything. And the 'New' Celestial God took the 'full-measure' of the cosmos, of where the Celestial Pole's position was at, during his climb's current era of Precession. Because the Young Upstart does this, looking to find the measure; he then starts the next World Age. And each Age was said to fall away into a flood, so that the next Age could arise and take its place. Each of the 12 Zodiac Ages ends with a flood as the new one takes its place.

It was also seen as the 12 Pillars of the Zodiac. The 'Falling Pillar' signaled the end of an old era and a new one taking its place. You can understand the metaphor easy. The Axis Moves... The pillar falls. And the Celestial Axis is in a new position.

The Templars discovered more than the words, visual images, and actions; to the Art of the Fugue. Perhaps the Muslim world still knew the Art of the Fugue, and they helped teach the Templars. Either way, the Templars were given a guide... the BIG PICTURE, to understand it.

Altogether, the puzzle pieces tell the story of humanity's past and our future, about the time table for Second Coming of Christ, and the evolution of each human's soul that is available to them... But only if they know how to unlock it.

....

Chapter 48: The Renaissance of the FUGUE

From the Templars in the 1200's and 1300's... to the Rosicrucians of the 1400's and the 1500's... to Masonics of the 1600's and 1700's... to the great Mystery Schools in the 1800's and 1900s', like The Hermetic Order of the Golden Dawn; they secretly passed down the Art of the Fugue. Dante Algieri, William Shakespeare, Lewis Carroll, J.R.R. Tolkien, Frank Herbert; and others more recent, writing books and making movies, using the Fugue. And they all got their knowledge from the Templar traditions. All this time, kept a secret. Disney and Hollywood used the Art from the very beginning of movie making. And music, Quinn the Eskimo, the 5th polar messiah.

And then came the incredible book, *Hamlet's Mill*, by Hertha von Dechend and Giorgio de Santillana; in 1968/69. And it blew the lid off the secret. And ever since then, anyone who reads this book is on a journey to understand this secret language. Is she really sure -all that glitters is gold, and just how is she buying -that stairway to heaven? The cat was out of the bag. Musicians, film makers, and writers of words, all over the world; had access to the Art of the Fugue.

....

Chapter 49: Recognizing the Puzzle Pieces

But the clues are not just as easy; as seeing a, b, and c; let alone spelled out like a + b = c. You can see this normal equation is all in order, the ducks all in their specific place, in the row. But for the Fugue... Clues are placed here there and everywhere, and not in correct order. You will see something the 1st time; and because you are not aware of anything about it, the first time it means nothing to you. Then after the 2nd read through, or the 3rd, you have seen enough of the other clues to really spot the hidden clues you missed before. These clues are meant to be mixed up and vague.

Years after Terry Gilliam's, the *12 Monkeys*, came out; I was watching the movie again, on cable. It was a good several years after I started studying the Fugue's Art. This movie is truly a masterpiece work; and my jack-in-the-box moment came. This is when one speech or one action stands out so much, POP like a jack-in-the-box, that it tells the reader/viewer... THIS IS AN ART OF THE FUGUE WORK. For *Jimmy Neutron*, it was when he held out the "mandible crest", the jawbone from Cindy's dinosaur. And I should have picked up on it sooner, since the very first scene has 'the Fall' just like a first scene should. For my jack-in-the-box moment in *12 Monkeys*, it's near the end of the movie. It's the scene in the movie theater where Dr. Reilly is putting disguises on Cole and herself; which is such a great framework of overt text.

I realized it was Art of the Fugue, when Cole gives a speech about his Fugue state. He's watching the movie at the theatre they are hiding in. Cole tells Katherine that he's confused because he can't remember things. James Cole tells Katherine Reilly… "What's happening to them is like what happens when you sit through a movie the second or third time around, it's different because your different. It's not that the movie changes, it can't change. But we have changed, we see different and new things." And As *Hamlet's Mill* (p.xi HM) says, "I had the answer there, but I was not ready to understand. This time I was able to grasp the idea at a glance, because I was ready for it."

My jack-in-the box-moment came early, in the Downey Jr. *Sherlock Holmes* movie. In the beginning sequence, after Holmes and Watson catch Lord Blackwood. The villain provokes Watson to run at him, and Holmes stops him from running into a glass rod at eye level, an axis. Watson is about to kill himself. After Holmes stops Watson's impalement; Watson asks Holmes, "How on Earth did you know that was there?" Holmes replies, "Because I knew to look for it". That's the Fugue! *If you don't know what to look for, you will never see it.* It's a see-through line, a glass rod, hidden between the lines.

By its very nature, the Fugue is trying to survive undetected, against hostile forces. So it's meant to be difficult to navigate. Truly its best defense is that nobody is every aware that there are secret meanings hidden inside.

At its heart, the Art of the Fugue is meant to be confusing, as a method of transmitting information. It's confusing to any who, would at a cursory level, try to make much sense of it. Thus, the Fugue is meant to have confusing clues. But because of that, the Fugue really does its best to explain how its own language works. It was called the Green Language and the Language of the Birds in the Romance Era. Yes, it is its own language, a language of symbols, actions and metaphors. Fortunately, all we need to do to understand the clues to the language of the Fugue, is simply to follow the thread with curiosity. Without the thread we would get lost. *Hamlet's Mill* is a guide that shows us the way; yet even with the guide, it's meant to be confusing, that's why it's called… the Fugue. Tongue-Vague, Fugue-Dialogue.

....

Chapter 50: The Qualities of a Pirate

Pirates are encoders, hijacking an overt text for the Fugue. Many movies have used the Art of the Fugue on some level. And some of the very best encoders of all time are the current ones, because movies can show so much visually.

While many movies discuss the various different Fugue topics; the topic about -the characteristics of the Fugue itself- and also 'how to encode' is best covered in the one of the wonderful Fugue movies by Disney's Ted Elliot and Terry Rossio. Yes *Aladdin* was Art of the Fugue; but it was the first *Pirates of the Caribbean: Curse of the Black Pearl*… that went out of its way to explain the qualities of the Fugue, and the concept of encoding. This movie teaches how to recognize encoding.

In the first scene of the movie, on the ship as the child Elizabeth Swan sings her pirate song. Mr. Gibbs sneaks up on her and stops her mid song…

<u>Gibbs</u> – "It's bad luck to be singing about pirates with us mired in this unnatural fog. Mark my words."

<u>Norrington</u> – "Consider them marked."

This very first scene in the movie begins the encoding about the Fugue characteristics. This specific line by Gibbs is a direct naming of the Fugue itself. "Unnatural" means 'man-made'; "fog" as in the 'Fugue'. And of course "fog" obscures sight, which lets things get hidden within it. The Phrase "mark my words" ties the Fugue to dialogue and writing. And the word "mired" specifically points to being stuck in the Fugue, causing confusion so that it remains hard to find; like the Fugue text hidden inside the overt text.

....

Mullroy and Murtogg, the fat and skinny sailors, ask Capt. Jack why he's there; demanding his honest answer. Sparrow says he's going to steal a ship for piracy. Murtogg chastises Capt. Jack for telling a lie; but Mullroy thinks Capt. Jack was being honest. Then Murtogg responds.

<u>Murtogg</u> – "If he were telling the truth, he wouldn't have told us."

<u>Capt. Jack</u> – "Unless he knew you wouldn't believe the truth, even if he told it to you."

Here we encounter the Amleth/Hamlet trait of never wanting to be telling a lie, so the truth that is told is so outlandish, or hidden in metaphor, that it seems a lie. Hamlet said that a knife which cuts ham, is a ship that cuts through the sea. And his guards recognized the cunningness in his craziness.

Jack speaks about commandeering. He's taking over something that already exists, for his own purpose, an overt text, especially since a ship is a vessel. It's a vessel for the fugue metaphors. Second, he says he's going to sail to Tortuga and pick up a new Pirate crew. A crew is what gets the work done, it's the tools that are used. Jack lists 4 criminal pirate actions. The movie uses the Fugue's 'doublespeak', which means picking the specific flavors of all your motifs and Fugue words. It's a Mast, not a tree or pillar. You can pick a theme, this movie uses pirating; *Ratatouille* uses cooking and spirits.

....

Capt. Jack cuts the corset off Ms. Swan with his knife.

<u>Mullroy</u> – "I never would have thought of that."

<u>Capt. Jack</u> – "Clearly you've never been to Singapore."

Very quickly in the text of *Hamlet's Mill*, we here about, <u>Heraclites</u> and his obscure utterances (p.v HM). The motif of obscure utterances is an example of how only those who understand the doublespeak get the extra meanings, but very few people understand the thing called the Green Language/the Language of the Birds/the Art of the Fugue. To the rest of you, who have never been to Singapore, you would never have thought of that.

....

Capt. Jack is in the Port Royal jail cell with the other captured pirates. Canon shots boom out in the night. Capt. Jack Sparrow tells the others that the ship firing on the port is none other than the infamous ghost ship, the Black Pearl. One of the captured pirates says that the ship and crew been pirating for almost 10 years and that there are never any survivors.

<u>Capt. Jack</u> – "No survivors? Then where do the stories come from I wonder?"

~The prisoners exchange confused (Fugued) expressions.

This explains how the Fugue's confusion leads to its own survival. In this instance The Black Pearl isn't just any ship (overt text), she's an encoded overt text. Another layer that I see, is how Disney's Ted and Terry might be bragging here, and very well deserved it is.

…Here's the doublespeak of bragging…

<u>Ted & Terry</u> – "We've made stories. We've been encoding into movies for near 10 years, and most of you never knew a thing."

~Welcome to the Art of the Fugue…

....

Wil Turner asks Capt. Jack where the Black Pearl has its berth, its home port. Sparrow is amazed that Wil has not heard the legend of the Black Pearl. He tells young Mr. Turner that the captain and his crew…

<u>Capt. Jack</u> – "…sail from the dreaded Isle de Muerta. It's an island that cannot be found, except by those who already know where it is."

First to note is the catch-22, with the silly and slightly insane nature of Jack's main verse. It sticks out! And because it sticks out, it makes the listener pause in their mind for a moment. It could be somebody's Jack-in-the-Box moment. It's just like the preceding passage, with no survivors, where do the stories came from. It made the other prisoner take pause to think. How wonderful the use of the word berth as a metaphor for birth; as in to create, meaning to encode a fugue production. In the passage above, the motif of the ship represents the overt text. So, 'making berth' means the encoding process itself. But the main player in this verse is the 'island'.

And just as the ship is the overt text, the island is the Art of the Fugue. The overt text is touched by the Fugue with motifs and topics… the ship touches the island where it makes berth.

Here's what the entire verse is actually saying in doublespeak…

<u>Will</u> – "You, Pirate!"

<u>The Fugue</u> – "Aye?"

<u>Will</u> – "You're familiar with the Art of the Fugue?"

<u>The Fugue</u> – "I know how it works-

<u>Will</u> – "How do I recognize it?"

<u>The Fugue</u> – "How do you recognize it? Don't you know any of the motifs and topics? the language of the birds and its motifs and topics cannot be understood by anyone unless they already know the meanings of the metaphors."

<u>Will</u> – "The Art of the Fugue is real enough, so there has to be a way to understand the hidden language that is encoded with it."

And there is another layer to the motif of the Isla de Muerta, that is 'not' about how to recognize the encoding in the Art of the Fugue. The 'Island of the Dead' refers to the most fantastic legends of the ancient Mystery Religions. They had secret knowledge how to evolve the human soul into ascension. The soul travels through the starry sphere. Lewis Carroll uses, -seeing the sun at midnight… It was a journey to realm of the divine, With the 'isle of the dead' as the gateway between our reality and the divine realm; just floating in outer space, which is the 'sea'. That's 2 different meanings for 2 different topics; how to encode and also soul ascension, all from the same verse.

……………

Wil Turner has just sprung Capt. Jack out of the jail cell. And the two have made it to the beach. They're looking at the port harbor and see the 2 ships of the Royal Navy. Will asks Jack if the two of them are planning to steal the bigger ship

<u>Will</u> – "We're going to steal a ship… that ship?"

<u>Capt. Jack</u> "Commandeer, we are going to commandeer… that ship. It's a nautical term.

Again the motif of the ship, it's the overt text again. We know this because we learn that it's going to be stolen, which means encoded -We learn that we are going to pass down information- we are finally given the clue that there will be a specific theme to the encoding. It will be given in a second language of nautical terms, so there will be motifs, of the sea, the ship, the sail, the anchor, the steering oar and mast.

....

Capt. Jack and Wil carry a small dinghy upside down for air while they walk under the sea. Wil steps into a lobster trap tied to a rope, that goes up to the surface, tied to a small floating barrel. It's A wonderful homage to Tolkien's 'Barrel Rider' motif from *The Hobbit*, which is about the soul ascension. It has nothing to do with the Precession. The dialogue is short and to the point…

Will – "This is either madness or brilliant."

Capt. Jack – "It's remarkable how often those two traits coincide."

Very Amleth/Hamlet… the insanity that is only a disguise for genius. This verse is the first of the next several that use the concept of 'the fool' or insanity as a linchpin for them being understood as Fugue speak.

....

In the bar at Tortuga Capt. Jack tells Gibbs his plan to take back the Black Pearl from Capt. Barbossa. Gibbs calls him a fool for even thinking he could do it, considering the legend of the curse. Capt. Jack insists he's not a fool, despite what Gibbs thinks.

Gibbs – "Prove me your wrong. What makes you think Barbossa will give up his ship to you?"

Capt. Jack – (pointing to Will) "Let's just say it's a matter of leverage aye. That is the child of Bootstrap Bill Turner, his only child."

Here we have a reference to the ship as the overt text. In it, Gibbs asks why Capt. Jack thinks that the overt text, (Capt. Barbossa) will let itself be encoded successfully (give up his ship to you). The response by Capt. Jack is that he knows how to successfully apply the Art of the Fugue (matter of leverage), so that it lasts through the generations (the child).

And it's all wrapped around the Amleth/Hamlet character trait of they who seems a fool, is only pretending to be a fool.

....

Gibbs has his new crew from Tortuga assembled for inspection by Capt. Jack and Wil. Gibbs asserts that the crew are all seasoned sailors; and also, crazy enough to go after the ghost ship.

Capt. Jack – "Mr. Cotton, do you have the courage and fortitude to follow orders and stay true in the face of danger and almost certain death?"

Gibbs explains to Capt. Jack and Wil, that this sailor had his tongue taken out. And he had a pet parrot that did the talking for him. Gibbs ends it with an admission of confusion (a Fugue trait) saying that nobody knows how Mr. Cotton trained the parrot to speak for him. So, Sparrow asks the parrot the same question. And then, Mr. Cotton's Parrot answers...

<u>Parrot</u> – "Aww, wind in the sails, wind in the sails."

<u>Gibbs</u> – "Mostly we figure that means yes."

<u>Capt. Jack</u> – "Of course it does, (to Wil) well are you satisfied?" -

<u>Will</u> – "Well you proved their mad."

This is the last in the train of references to the fool/madman. There are other mentions in the script, to be sure; but these are the important ones that stand out beyond just being a simple reference to insanity. Here, Capt. Jack is asking Mr. Cotton, (the overt text), a question and getting told an answer in the previously established second language of Nautical Terms. What a brilliant little piece about how the encoding process gets recognized by those seeking to find it. The parrot is the Fugue, riding on the overt text.

··

Chapter 51: The Storyteller's Tale

On page 10 of *Hamlet's Mill*, the 2 esteemed authors, Giorgio de Santillana and Hertha von Dechend, share a famous quote about Isaac Newton by John Maynard Keynes. I added bold italics to certain phrases, and increased the size of words where it is most relevant.

> "Newton was not the first of the Age of Reason. He was the last of the magicians, the last of the Babylonians and Sumerians, the last great mind which looked out on the visible and intellectual world with the same eyes as those who began to build our intellectual world rather less than 10,000 years ago. . . Why do I call him a magician? **Because he looked on the whole universe and all that is in it as a riddle, as a secret** which could be read by applying pure thought to certain evidence, **certain mystic clues which God had laid about** the world to allow a sort of philosopher's treasure hunt to the esoteric brotherhood. He believed that **these clues were to be found partly in the evidence of the heavens** and in the constitution of elements (and that is what gives the false suggestion of his being an experimental natural philosopher), **but also partly in certain papers and traditions handed down by the brethren in an unbroken chain back to the original cryptic revelation in Babylonia.** He regarded the universe as a cryptogram set by the Almighty-just as he himself wrapt the discovery of the calculus in a cryptogram when he communicated with Leibniz. By pure thought, by concentration of mind, the riddle, he believed, would be revealed to the initiate ["Newton the Man," in The Royal Society. Newton Tercentenary Celebrations (1947), p. 29.]."

Keynes is famous as a philosopher and economist. But he also became an authority on Sir Isaac when he bought a collection of his papers that had been found. This quote was from the speech *Newton, the Man*, that Keynes brother gave for Keynes post-humously; at the tercentenary celebration of Sir Isaac Newton's birth in 1946.

The 'brethren' is a name that the Free Masons call each other; like the 'son of a widow'. And Free Masonry came from the Rosicrucian traditions; which itself came from the Templars. And so here we have, what is in my opinion, the most important thing about this entire quote. The Free Masons have papers from scrolls dug up underneath the Temple Mount in Jerusalem, that trace a history back to some cryptic revelation in ancient Babylon; probably the Jewish Captivity, 500 years before Christ... What's this Famous quote mean? To me it means, some high muckity-muck Free Mason is bragging that they have a secret history and knowledge that goes back to 500 BC. Obviously not every Free Mason is like the potentate John Maynard Keynes, and in on the secret knowledge. But it is my belief that Masonry in the last two centuries has taken in great writers among their ranks and given them the secret knowledge; so they could use the Art of the Fugue, as they create their stories to entertain.

And that is how Masonic authors like John Ronald Reuel Tolkien learned the Art of the Fugue from the Templar lore; before *Hamlet's Mill* came out. Three of my other favorite Fugue authors in Sci-Fi / Sword and Sorcery are, Frank Herbert, Michael Moorecock, & Roger Zelazny. And from the century before is the poem, *The Walrus and the Carpenter*, in *Through the Looking Glass*, by Lewis Carroll. It's is a brilliant example of a poem which is so short; but still covers so many topics in the Art of the Fugue. Modern additions to the endless library of the Fugue's Art are *Skinny Legs and All* by Tom Robbins; and *The DaVinci Code*, by Dan Brown, which is a 1st class masterpiece.

The Fugue has been in movies since movies began. The French film in 10 parts, *Les Vampires* from 1915-1916 is a masterpiece. And especially in the last several decades, the Fugue movies have exploded. Some of the great writers are listed here, although not in any specific order, Brad Bird, Ted Rossio & Terry Elliot, Terry Gilliam, Alex Proyas, Guillermo del Toro, J.J Abrams, Luc Besson, even Bruce Willis for *Hudson Hawk*; especially Bruce Willis for writing *Hudson Hawk*, it's a masterpiece. The list of names for the Art of the Fugue writers in Hollywood goes on. Peter Weir, Scott Rudin, Stephen Sommers, Len Wiseman, Andrew Nichols, Tim Burton. Also, some great Fugue movies to name are; *Jimmy Neutron, National Treasure, Chronicles of Riddick, Highlander, Life Force, Galaxy Quest, Sherlock Holmes* (Ritchie), *Watchmen, Megamind, Tangled, Lara Croft, Harry Potter, The Librarian, The Matrix, Evil Dead III, ...*

I have never met these writers and directors, and so I do not say these people have ever 'claimed' to use the Art of the Fugue. But several authors in Precessional Cosmology do mention it, especially John Major Jenkins and Graham Hancock in various books and web sights of theirs.

Chapter 52: The Elephant in the Room

And now finally... The modern-day ILLUMINATI.

In my opinion, there are several groups that combine under that umbrella term – the Children of the Light-. And I don't believe they all work together in harmony, but are thrown together by necessity.

The movie Hudson Hawk 1991, from Bruce Willis, did a 'fair' job of pointing them out.

The Bilderbergers, or by whatever name they are called, are comprised of the one hundred richest people in the world. Whoever they are at the time.

The movie has a British butler who wields blades. For me, that's good enough to represent the Anglo-American SION, who are descended from the sword wielding Templar Knights

The Vatican is implicated and so are the heads of organized crime, but several groups throughout the civilized world, not just the Sicilian / Mafia.

The CIA in the Movie represent two different types of subgroups. The first and obvious are the shadow organizations of governments, certainly the CIA itself, NSA, MI-6, etc... as information gathering bodies, but also such organizations like the Bohemian Grove Group, that is said to recruit its ranks from select / old money members of Scottish Rite Masonry.

The second subgroup, represented by the CIA in the movie, are named by their leader... The MTV-I-A's. It is public knowledge that the Bilderbergers meet, now it is said, twice a year, for a three-day conference. They summon to their side the world's leaders / politicians. And also, they invite some world class journalists; as well as, some of the leading Pop, Rock, and Rap superstars. Thus, the MTV-I-A's. In urban circles they are called the -Boule- which, now means the Black Illuminati of America.

And of course, for our generation's conspiracy theories, there is the Hollywood Illuminati, led by Disney.

Finally, the thieves, they are the pirates, those who understand the Art of the Fugue and use it. Thieves because it's not ours by origin, we have only taken it from all those who used it before us. And we steal another story to insert ours into.

The Art of the Fugue is now in movies, books, songs, and music videos. The Illuminati is a relatively new name, just a few centuries old and certainly not the same then as now. When the ancient Greeks were using this Art, they didn't call themselves that, and they weren't planning on taking over the world. Except of course, for Alexander the Great. As I just pointed out, the Illuminati did not create the Fugue, and it does not belong to them. They just use it. Anyone can.

....

Open the way, I will take the path!
The fixed times have declined.
The set day has passed.

Appendix: The Science of Astronomy

For technical support, I must acknowledge and graciously thank Dr. Edwin C. Krupp of the Griffith Observatory in Los Angeles, California. As I am a total non-expert in the field of astronomy, I thought it was best to consult a world-class expert on some of my questions. I will always be grateful for his willingness to answer me, as valuable as his time must be.

By Email correspondence, he has informed me...

-1) THE RATE OF PRECESSION: With modern measurements, it shows precession does not keep a constant pace. In the year 2000 "the value is actually 71.583433", but since then it has changed (17 July 2017). And also, "No serious expert in celestial dynamics quotes precession with the precision you mentioned {25920 years}. The most authoritative references simple state the cycle is "about 26000 years."... The Platonic Great Year is a concept from antiquity, not a period linked to modern understanding of solar system dynamics... This is an historical issue {25,920}, not an astronomical issue." (21 April 2008)

-2) According to the Dr. Krupp, "The current estimate of the {Celestial Axis and its} obliquity of the Ecliptic {Axis} is 23°26'21.4110". (21 April 2008) Several different sources place the obliquity between the Celestial and Ecliptic Axis', currently at: 23.4 degrees, 23.44 degrees, & 23.5 degrees.

-3) As far as the position of the Celestial Poles and any movements they make are concerned, Dr. Krupp had this to say... "There is a variety of smaller scale motions with cyclical and secular variations that affect the exact location of the poles. Whether they are of any consequence depends on the point you are trying to make. For most issues, they are not consequential. For some, they are." (21 April 2008 & 17 July 2017)

-4) Dr. Krupp explained the complete cycle of 21,000 years for the rotating dates for the distance between the sun and Earth being closest and farthest from each other. This cycle is actually thought to be a combination of a 23,000-year cycle and a 19,000-year cycle. And also how it moves with the solar orbit direction (proceeds) and not against it like Precession does (21 April 2008)

-5) Dr.Krupp confirmed my calculation that, "If you map the position of the north galactic pole accurately enough, it falls on a diameter that will be oriented north/south (will coincide with the local celestial meridian) on 3 April." (10 November 2017)

....

I'm sure that professional astronomers will find fault with my 'generalized' facts. For instance, the Earth's orbit is not exactly the Ecliptic Plane. The Ecliptic Plane is the Sun's apparent motion through the starry sphere from our observation here on Earth.

Obviously my inexpertise in astronomy left me with many questions, some of which I could find answers for; either online, or with me attempts to communicate with professional astronomers. My prime example is... Why does the wobble of the Ecliptic Plane through the Invariable Plane have specific intersection dates, currently at Jan.9th and July 9th; just like the specific dates for the intersection of the Solstice and the historic Galactic Center; But the intersection of the Galactic Plane and the Celestial Plane does not have specific intersection dates, and can only be determined by the intersection's alignment with the north/south Celestial meridian at noon and midnight? And this one was answered by Dr. Krupp. "It doesn't make any sense to say the galactic equator crosses the celestial equator on two specific dates. An intersection with the celestial equator is not affiliated with any specific date because dates are not assigned to the celestial equator." (10 November 2017) Some questions still remain unanswered. My main unanswered question is this. Is my calculation correct that the Galactic and Celestial Equators are aligned with the Celestial Meridian on the dates of Jan.1st and July 3rd?

I give an open invitation to all, please prove or disprove that...
 -1) my diagram of the 2/7^{ths} overlaid on top of my so-called Masonic Cross (using the correct 1/7th angle and not the slightly incorrect 13 by 27 checkerboard), by using proofs; does create the perimeter of the Circled-Square and the volume of the Cubed-Sphere.
 -2) The nexus' of the Galactic and Celestial Equators align with the Celestial Meridian on January 1st and July 3rd.
 -3) The Ecliptic Axis shares a parallel plane with the Galactic Axis, like the 2 hands of a clock at 10 and 12 that share the same plane of the face of the clock.
 -4) The Great Pyramid's airshafts are at the angle of 2/7^{ths}.

The first who does so, that can be verified, will have their proof added to this Appendix and be given credit for their work, even if they cite another source. And of course, I will appreciate anyone who points out corrections that need to be made in my text. So please let me know.

....

Again, I state- *Hamlet's Mill* is considered controversial by many. Reader's either love it or hate it. Graham Hancock has defended it from detractors on his website. The haters say it is a confusing jumble of ideas. As I have shown, it was meant to be as a mechanism of survival through the ages. The detractors say it has no validity. In the opinion of this mythographer, the literary evidence is there to support their claims.

Gauntlet throne!

Diagrams originally from other sources:

....

*These figures are all in the Public Commons. However, I first encountered them in the book;

Mapping of the Heavens, Peter Whitfield @ 1995
First published 1995 by the British Library
Published in North and South America by
Pomegranate Artbooks, Box 6099,
Robert Park, California 94927
ISBN 0-87654-457-8
Pomegranate Catalog No. A803

....

Public Commons, Backer's Star Chart, c. 1710.
The Library of Congress, Washington D.C.

Fig. 30: *Backer's Star Chart, Bernice's Hair #1 .... p. 30

Fig. 31: *Backer's Star Chart, Bernice's Hair #2 .... p. 30

Fig. 35: *Backer's Star Chart, the Serpent Bearer #3 .... p. 33

Fig. 37: *Backer's Star Chart, the Man-Child #1 .... p. 35

....

Public Commons, Copernicum System:
Astronomer's own Diagram from
The original manuscript (Fascimile)
The British Library, X0622/170

Fig. 70: *Copernicus' Manuscript # 1 .... p. 60

Fig. 71: *Copernicus' Sun-Centered System #1 .... p. 61

....

Public Commons, Aratus's Map of the Heavens,
From a 15th Century Manuscript
The British Library, Add. MS 15819, f3

Fig. 41: *Aratus's Star Chart, Gemini and Ophiuchus #1 .... p. 39

Fig. 42: *Aratus's Star Chart, Gemini and Ophiuchus #2 .... p. 40

....

Public Commons, Hevellius's Star Chart 1690
From *Uranographia*,
The British Library, 532.k.19.1

Fig. 39: *Hevellius's Star Chart, the Shield #1 .... p. 38

Fig. 51: *Hevellius's Star Chart, the Rod of Iron #3 .... p. 45

Public Commons, detail from the Caprarola Freso,
1575 By Anonymous Artist in
the Villa Farnese at Caprarola
Photograph: Scala

Fig. 24: *Caprarola Fresco, The Hydra #1 .... p. 24

Fig. 34: *Caprarola Fresco, the Serpent Bearer #2 .... p. 33

Fig. 38: *Caprarola Fresco, the Man-Child #2 .... p. 36

Public Commons, anonymous Medieval Star Chart

Fig. 1: *Medieval Star Chart, North Ecliptic Pole #1 .... p. 8

Fig. 2: *Medieval Star Chart, North Celestial Pole #1 .... p. 8

Fig. 55: *Medieval Star Chart, North Celestial Pole #1 .... p. 51

Fig. 62: *Medieval Star Chart, North Cel. & North Ecl. Pole #1 .... p. 55

Public Commons, Apian's Star Chart
from *Astronomicon Caesareum*, 1540
The British Library, Maps C.6.d.5

Fig. 27: *Apian's Star Chart, Virgo #1 .... p. 28

Fig. 28: *Apian's Star Chart, Virgo #2 .... p. 29

Fig. 33: *Apian's Star Chart, the Serpent Bearer #1 .... p. 33

Fig. 36: *Apian's Star Chart, the Serpent's Tail and the Eagle's Wing #1 .... p. 34

Fig. 50: *Apian's Star Chart, the Rod of Iron #2 .... p. 44

Public Commons, Honter's Star Chart, 1541
The British Library. Maps C.6.d.5

Fig. 29: *Honter's Star Chart, Virgo #3 .... p. 29

Public Commons
F. Argelander, *Uranometria Nova*,
1843 The British Library, 14000.c.42

Fig. 32: *Argelander's Star Chart, Bernice's Hair #1 .... p. 31

Fig. 40: *Argelander's Star Chart, the Shield #2 .... p. 38

+These figures have been altered at least 3 ways from the sources I found them in;

Taken and altered from;
Norton's 2000.0 Star Atlas and Reference Handbook,
edited by Ian Ridpath, Eighteenth Edition
copyright Longman Group UK Limited 1989
Longman Scientific & Technical
Longman Group UK Limited
Longman House, Burnt Mill, Harlow
Essex CM202JE, England
Copublished in the United States with
John Wiley & Sons Inc., 605 third Avenue, new York, NY 10158

Fig. 25: +the Hydra #2 .... p. 25

Fig. 43: +Northern Hemisphere #1 .... p. 41

Fig. 44: +Northern Hemisphere #2 .... p. 41

Fig. 52: +The Rod of Iron #4 .... p. 46

Fig. 112: +The Galactic & Celestial Equators #1 .... p. 81

Fig. 124: +The Galactic & Celestial Equators #1 .... p. 88

Fig. 126: +The Galactic & Celestial Equators #1 .... p. 89

Fig. 129: +The Galactic & Celestial Equators #1 .... p. 90

Taken and altered from; *Maya Cosmogenesis 2012*
copyright 1998 by John Major Jenkins
Bear & Company Inc, Publishing
Santa Fe, NM 87504-2860

Fig. 12: +Cylinder Proj., Ecl. Armillary w- Gal. Equator #1 .... p. 13

Fig. 13: +Cylinder Proj., Ecl. Armillary w- Gal. Equator #2 .... p. 14

Fig. 47: +Cylinder Proj., Ecl. Armillary w- Gal. Equator #1 .... p. 43

Fig. 48: +Cylinder Proj., Ecl. Armillary w- Gal. Equator #2 .... p. 43

Fig. 49: +Cylinder Proj., the Rod of Iron #1 .... p. 44

....

Taken and altered from; *The Message of the Sphinx*.
Graham Hancock & Robert Bauval.
Published by Three Rivers Press, 201 East 50th Street, New York, New York 10022.
Member of the Crown Publishing Group.

Fig. 65: +Giza Pyramid & the 1/7th Angle #1 .... p. 58

Fig. 66: +Giza Pyramid & the Airshafts # 1 .... p. 58

....

Taken and altered from;
https://www.google.com/earth. Chichén Itzá, Tinum, Mexico

Fig. 74: +Chichen Itza, Archeological Site #1 .... p. 62

Fig. 75: +Chichen Itza, Archeological Site #2 .... p. 63

Fig. 76: +Chichen Itza, Archeological Site #3 .... p. 63

Fig. 77: +Chichen Itza, Archeological Site #4 .... p. 63

Fig. 79: +Chichen Itza, The Plaza #1 .... p. 64

....

Taken and altered from; *Hamlet's Mill*
copyright 1969 by Giorgio de Santillana and H. von Dechend
First paperback edition published in 1977 by
David R. Godine, Publisher, Inc.
Box 450 Jaffrey, New Hampshire 03452
a Nonpareil Book

Fig. 54: +North Celestial & Ecliptic Poles #2 .... p. 51

Public Commons
William Cunningham, *The Cosmographicall Glasse,* London 1559
Used in, The Armillary Sphere in Poetry, Literature, and Art.
http://www.sites.hps.cam.ac.uk/starry/armillpoems.html
Copyright 1999, Adam Mosley
and the Department of History and Philosophy of Science
of the University of Cambridge

Fig. 100: +Atlas Holds the Heaven's #1 .... p. 74

Public Commons Hevellius's Star Chart 1690
From *Uranographia*, Via Pinterest

Fig. 26: +Hevellius' Star Chart, Draco #1 .... p. 26

Leonid meteor shower best views before dawn November 17, 2014
https://phys.org/news/2014-11-leonid-meteor-shower-views-dawn.html
Provided by: University of Texas at Austin
Article by Rebecca Johnson. November 14, 2014
Taken and Altered from a figure originally from
stardate.org, Credit: The University of Texas at Austin

Fig. 14: +Leonids Meteor Shower #1 .... p. 15

A depiction of the 1866 Leonid Meteor storm, produced in 1889
for the **Seventh-day Adventist** book *Bible Readings for the Home Circle.*
Taken from the Public Commons and Wikipedia
https://en.wikipedia.org/wiki/Leonids

Fig. 15: +Leonids Meteor Shower #2 .... p. 16

Taken and Altered from,
The Queen Mary Apocalypse. Copyright, Lessing Archive. From
http://tanjand.livejournal.com/1072178.html
and also, https://www.pinterest.com/pin/527765650057065018/

Fig. 63: +The Dragon and the Beast #1 .... p. 56

Fig. 121: +The Dragon and the Beast #1 .... p. 87

Public Commons, taken and altered from,
The Zodiac from John P. Pratt Home Page

Fig. 18: +Zodiac Constellations and Signs #1 .... p. 19

....

Taken and altered from, *Chichen Itza;*
Astronomical Light and Shadow Phenomena of the Great Pyramid
Lic. Miguel Angel Vergara C.

Fig. 73: +Chichen Itza, Sun Rise & Sun Set #1 .... p. 62

....

Public Commons, taken and altered from,
Wikimedia Commons... File:Stonehenge Distance.jpg
Nedarb at English Wikipedia. Nov. 27, 2005,
from a photo Nov. 26, 2005

Fig. 80: +Stonehenge #1 .... p. 64

....

Public Commons, taken and altered from,
Wikimedia Commons...
File:US Navy 030926-F-2828D-307
Aerial view of the Washington Monument.jpg
U.S. Navy Photo, Sept. 26, 2003

Fig. 80: +Washington DC #1 .... p.64

....

Public Commons, taken and altered from,
Wikimedia Commons...
NASA, Mysid – vectorized by Mysid in Inkscape
http://earthobservatory.nasa.gov/Librry/Giants/
Milankovitch/Images/obliquity.jpg

Fig. 108: The Full Celestial Tilt Cycle #1 .... p. 79

....

xvii

References & Bibliography

(and page number where the quote, reference, or summary is found in this text)

....

Revelations of Saint John the Divine, The New Testament

p.3, 6, 7, 14, 16, 17, 24, 26, 31, 39, 40, 41, 47, 50, 53, 55, 56, 70, 77, 87, 92

...

The Book of Ezekiel, The Old Testament

p.17

...

The Book of Judges. The Old Testament

p.42

...

The Book of Genesis, The Old Testament

p.71, 92

...

Maya Cosmogenesis 2012, John Major Jenkins

Bear & Company; 1998; ISBN-13: 978-1879181489

p.14, 27, 76

...

Maya Cosmos, *David Freidel*, Linda Schele, and Joy Parker

William Morrow Paperbacks; 1993; ISBN-13: 978-0688140694

p.27,

...

Hamlet's Mill, Hertha von Dechend & Gorgio de Santillana

Nonpariel books; 1969; ISBN: 0-87923-215-3

p.4, 14, 47, 94, 95, 96, 98, 99, 100, 101, 105, 106

...

Inner Reaches of Outer Space; Joseph Campbell
Harper and Row; 1986; ISBN: 1-57731-209-0
p. 47

...

Grimnismol, Poetic Edda; Snorri Sturluson, Trans. Henry Adams Bellows
Princeton University Press; 1936
http://www.sacred-texts.com/neu/poe/poe06.htm
p.48

...

Heaven's Mirror; Graham Hancock, Santha Faiia
Three Rivers Press; 1998
ISBN-10: 0609804773
p.48

...

The Secret Architecture of Our Nation's Capital, David Ovason
Harper perennial; reprint edition; 2002
ISBN-13 978-0060953683
p.64, 92

...

Death of the Gods in Ancient Egypt; Jane B. Sellers
Lulu.com; 2007; ISBN-10: 140317906
p.49

...

The Egyptian Book of the Dead; E.A. Wallis Budge
Dover Publications; modern reprint edition; 1967
ISBN-13: 978-0486218663
p.48

...

The Celestial Ship of the North, Vol. II; E. Valentine Straiton
D.G. Nelson Publisher; 1927; ASIN: B00086MUIQ
p.51

...

Fingerprints of the Gods; Graham Hancock

Three Rivers Press; 1995; ISBN-10: 978051788

p.47, 48

...

Message of the Sphinx; Graham Hancock, Robert Bauval

Broadway Books; 1 edition; 1997; ISBN-13: 978-0517888520

p.48, 58

...

Chichen Itza; Astronomical Light and Shadow Phenomena of the Great Pyramid; Miguel Angel Vergara C.

p.61

...

Apokatastasis, Chapter 7 – Cycles; July 2017

http://english.boek-apokatastasis.nl/?page_id=173

p.80

...

Earth Orbit Inclination to Invariable Plane; July 2017

https://www.vcalc.com/wiki/Juliet/Earth+orbit+inclination+to+invariable+plane

p.80

...

Uriel's Machine; Robert Lomas, Christopher Knight

Random House; 1999; ISBN: 97870712680073

p.64

...

Inanna, Queen of Heaven and Earth;

Diane Wolkstein, Samuel Noah Kramer

Harper Perennial; 1st edition; 1983; ISBN-13: 979-0060908546

p.95

...

Enuma Elish; Leonard W. King
Book Tree; 1999; ISBN-13: 978-1585090419
p.95

...

Don Quixote; Miguel de Cervantes, Trans. Edith Grossman
Harper Perennial; reprint edition; 2005; ISBN-13: 978-0060934347
p.95

...

The Odyssey; Homer, Trans. Robert Fagles
Penguin Classics; deluxe edition, 1999; ISBN-13: 978-0140268867
p.96

...

Homer's Secret Iliad; Florence Wood, Kenneth Wood
John Murray Publishers Ltd; first edition; ISBN-13: 978-0719557804
p.96

...

Skinny Legs and All; Tom Robbins
Bantam Books; 1990; ISBN 0-553-05775-8
p.106

...

The DaVinci Code; Dan Brown
Doubleday (US); 2003; ISBN 0-385-50420-9 (US)
p.97, 106

...

The Walrus and the Carpenter, Through the Looking Glass; Lewis Carroll
Create Space International Publishing Platform; 2017
ISBN-13: 978-1546415939
p.103, 106

...

The Hobbit; John Ronald Reuel Tolkien

George Allen & Unwin (UK); 1937

p.104

...

Hamlet; William Shakespeare

Penguin Classics; 2016; ISBN-13: 978-0143128540

p.93, 101, 104

...

Newton the Man; John Maynard Keynes;

The Royal Society. Newton Tercentenary Celebrations; 1947

p.105-106

....

Personal emails; Dr. Edwin C. Krupp;

2008, 2016, & 2017

p.83-84

....

Aquarius/Let the Sunshine, Hair; James Rado & Gerome Ragni,

The 5th Dimension, Soul City; 1969;

p.19

....

Lost; Edward Kitsis and Adam Horowitz

Dist: ABC; 2004-2010   Prod: Bad Robot Productions, ABC Studios

p.48, 106

...

Once Upon a Time; Edward Kitsis and Adam Horowitz

Dist: ABC; 2011-current   Prod: ABC Studios, Kitsis-Horowitz

p.48, 106

...

The Librarian; David Titcher

Dir: Peter Winther Prod: Dean Delvin. Electric Entertainment

Dist. TNT; 2004

p.106

....

The Fifth Element: Luc Besson and Robert Mark Kamen

Dir: Luc Besson Prod: Gaumont

Dist: Gaumont, Buena Vista, International, Columbia Pictures; 1997

p.48, 106

...

National Treasure; Jim Kouf, Cormac Wibberley, Marianne Wibberley

Dir: Jon Turteltaud

Prod: Walt Disney Pictures, Jerry Bruckheimer Films, Junction Entertainment, Saturn Films.

Dist: Walt Disney Studios Motion Pictures; 2004

p.49, 106

...

Pirates of the Caribbean I; Ted Elliot, Terry Rossio

Dir: Gore Verbinsky

Prod: Walt Disney Pictures, Jerry Bruckheimer Films

Dist: Buena Vista Pictures Distribution; 2003

p.93, 94, 100, 101, 102, 103, 104, 105

...

Galaxy Quest; David Howard, Robert Gordon

Dir: Dean Parisot

Prod: DreamWorks SKG, Gran Via Productions

Dist: DreamWorks Pictures; 1999

p.96, 106

...

The DaVinci Code; Akiva Goldsman

Dir: Ron Howard

Prod: Image Entertainment, Rainmaker Digital Effects, Skylark Productions.

Dist: Columbia Pictures; 2006

p.97

...

12 Monkeys; David Peoples, Janet Peoples

Dir: Terry Gilliam   Prod: Atlas Entertainment Classico

Dist. Universal Pictures; 1995

p.99

...

Jimmy Neutron;

John A. Davis, Steve Oedekerk, J. David Stem, David N. Weiss

Dir: John A. Davis

Prod: Nickelodeon Movies, O Entertainment, DNA Productions

Dist: Paramount Pictures; 2001

p.99, 106

...

Sherlock Holmes; Michael Robert Johnson, Anthony Peckham, Simon Kinberg

Dir: Guy Ritchie

Prod: Silver Pictures, Village Roadshow Pictures, Wigram Productions, Alcon Entertainment.

Dist: Warner Bros. Pictures, Roadshow Entertainment, 2009

p.99, 106

...

Aladdin; Ron Clements, John Musker, Ted Eliot, Terry Rossio

Dir: Ron Clements, John Musker

Prod: Walt Disney Pictures, Walt Disney Feature Animation.

Dist: Buena Vista Pictures; 1992

p.100

...

Les Vampires; Louis Feuillade
Dir: Louis Feuillade   Prod: Gaumont
Dist: Gaumont; 1915-1916
p. 106

...

Hudson Hawk; Bruce Willis, Robert Kraft
Dir: Micheal Lehmann   Prod: Silver Pictures
Dist. TriStar Pictures; 1991
p. 91 & 106

...

Chronicles of Riddick; David Twohy
Dir: David Twohy   Prod: Radar Pictures, One Race Films
Dist: Universal Pictures; 2004
p.106

...

Highlander; Gregory Widen, Peter Bellwood, Larry Ferguson
Dir: Russel Mulcahy
Prod: Cannon Films, Highlander Productions Limited.
Dist: Emi Films (UK), 20th Century Fox (US); 1986
p.106

...

Life Force; Dan O'Bannon, Don Jakoby
Dir: Tobe Hooper   Prod: Cannon Films
Dist: TriStar Pictures; 1985
p.106

...

Watchmen; David Hayter, Alex Tse
Dir: Zack Snyder
Prod: Legendary Pictures, DC Comics, Lawrence Gordon Productions
Dist: Warner Bros. Pictures (North America). Paramount Pictures (international); 2009
p.106

...

Megamind; Alan J. Schoolcraft, Brent Simmons, Guillermo Del Toro

Dir: Tom McGrath

Prod: DreamWorks Animation, Pacific Data Images

Dist: Paramount Pictures; 2010

p.106

...

Tangled; Dan Fogelman

Dir: Nathan Greno, Byron Howard

Prod: Walt Disney Pictures, Walt Disney Animation Studios

Dist: Walt Disney Studios Motion Pictures; 2010

p.106

...

Lara Croft; Patrick Masset, John Zinman

Dir: Simon West

Prod: Mutual Film Company, BBC, Lawrence Gordon Productions, Marubeni Corportation, Eidos Entertainment, KFP Production GmbH & Co. KG

Dist. Paramount Pictures; 2001

p.106

...

Harry Potter; Steve Kloves

Prod: Heyday Films, 1942 Pictures, Duncan Henderson Productions

Dist: Warner bros. Pictures; 2001

p.106

...

The Matrix; Lana Wachowski, Lilly Wachowski

Dir: Lana Wachowski and Lilly Wachowski

Prod: Village Roadshow Pictures, Groucho II Film Partnership, Silver Pictures

Dist: Warner Bros. (US); 1999

p.106

...

Evil Dead III, Army of Darkness; Sam Raimi and Ivan Raimi
Dir: Sam Raimi
Prod: Dino De Laurentis Communications, Renaissance Pictures
Dist: Universal Pictures; 1992
p.106

....

www.ingramcontent.com/pod-product-compliance
Lightning Source LLC
Chambersburg PA
CBHW051018180526
45172CB00002B/395